About Island Press

Since 1984, the nonprofit Island Press has been stimulating, shaping, and communicating the ideas that are essential for solving environmental problems worldwide. With more than 800 titles in print and some 40 new releases each year, we are the nation's leading publisher on environmental issues. We identify innovative thinkers and emerging trends in the environmental field. We work with world-renowned experts and authors to develop cross-disciplinary solutions to environmental challenges.

Island Press designs and implements coordinated book publication campaigns in order to communicate our critical messages in print, in person, and online using the latest technologies, programs, and the media. Our goal: to reach targeted audiences—scientists, policymakers, environmental advocates, the media, and concerned citizens—who can and will take action to protect the plants and animals that enrich our world, the ecosystems we need to survive, the water we drink, and the air we breathe.

Island Press gratefully acknowledges the support of its work by the Agua Fund, Inc., The Margaret A. Cargill Foundation, Betsy and Jesse Fink Foundation, The William and Flora Hewlett Foundation, The Kresge Foundation, The Forrest and Frances Lattner Foundation, The Andrew W. Mellon Foundation, The Curtis and Edith Munson Foundation, The Overbrook Foundation, The David and Lucile Packard Foundation, The Summit Foundation, Trust for Architectural Easements, The Winslow Foundation, and other generous donors.

Creating Green Roadways

James L. Sipes and Matthew L. Sipes

Creating Green Roadways

Integrating Cultural, Natural, and Visual Resources into Transportation

James L. Sipes and Matthew L. Sipes

Washington | Covelo | London

Library of Congress Cataloging-in-Publication Data
Sipes, James L.
Creating green roadways : integrating cultural, natural, and visual resources into transportation / James L. Sipes and Matthew L. Sipes.
p. cm.
Includes bibliographical references and index.
ISBN 978-1-61091-358-4 (cloth : alk. paper)—ISBN 1-61091-358-2 (cloth : alk. paper) – ISBN 978-1-61091-375-1 (pbk. : alk. paper) – ISBN 1-61091-375-2 (pbk. : alk. paper) 1. Roads—Design and construction—Human factors. 2. Context sensitive solutions (Transportation) 3. Highway planning. 4. Ecological landscape design. 5. Traffic safety. I. Sipes, Matthew L. II. Title.
TE175.S52 2012
625.7028'6–dc23
2012022226

Printed using Times New Roman
Typesetting by Blue Heron Typesetting

 Printed on recycled, acid-free paper

Manufactured in the United States of America
10 9 8 7 6 5 4 3 2 1

Keywords: asphalt, bioswale, bridges, climate change, concrete, construction, context sensitive design, Department of Transportation, highway infrastructure, land use planning, native plants, NEPA, public transportation, right of way, road diet, road ecology, safety standards, scenic byways, smart growth, traffic calming, trails, wildlife habitat

To Granny

and special thanks to the Sipes clan
for their support and encouragement

CONTENTS

ACKNOWLEDGMENTS

First, thanks to all of the firms, agencies, and individuals that helped put this book together, provided input, and shared photographs of their projects. To Heather, Courtney, and the editors at Island Press, thanks for making the publishing process only semi-painful.

To Grant Jones, you have been an inspiration over the years. Without the opportunity to work with Richard Shaw and the talented team at Design Workshop, I would still be trying to understand the concepts of green roadway design. To Laura Lewis, Cherokee National Forest, thanks for inspiring me to write a book like this so that people like you will have help in protecting natural resources while developing environmentally sensitive transportation projects.

Matt, thanks for writing this book with me. Thanks to Karen Collins for developing the book cover. Special thanks to Angie Sipes, who handled all of the little details for this book and is truly the queen of captions. I can't imagine doing a book without her help.

—James Sipes

To start, I thank the engineering faculty at the University of Kentucky, who taught me the basics of transportation design. During my studies, I thought it was ironic that we actually studied several of the green road projects my dad worked on, several of which are included in this book. It has been exciting to be able to apply green road principles to my work, and I look forward to continuing to explore how we can meet transportation needs while being respectful of the world around us.

I also thank HMB Professional Engineers, Inc., for giving me an opportunity to continue to develop as a professional engineer and for allowing me to learn from some of the best and brightest. And finally, thanks to Bekah for supporting me, offering encouragement, and even helping put this book together.

—Matthew Sipes

Chapter 1

Introduction

A road . . . is just a road.

It is a ribbon of asphalt, striping, and guard rails that allow automobiles to get from one point to another.

But a road can be something more.

A road can not only meet transportation needs, but it can also be a community asset that is as much a part of what defines a place as parks, houses, trails, and sidewalks.

A road can be respectful of people and place. The footprint of the road can be minimized to prevent negative impacts on cultural and natural resources, and the design of the road can ensure that existing resources are protected. In rural areas, roads can be an integral part of the landscape; the Blue Ridge Parkway shows us how it can be done. In any landscape, a green roadway should have a positive influence on cultural, environmental, and economic sustainability.

Creating Green Roadways is about this idea of a road being more than just a road. With creativity, sensitivity, passion, and technological know-how, we can build roads that are not only safe, practical, and buildable, but also environmentally friendly, visually attractive, and helpful in creating a sense of community.

This book advocates a way of approaching road design that integrates cultural, natural, and visual resources into the transportation design and planning process in an economically viable and socially significant way.

A green roadway is energy efficient, addresses long-term concerns such as climate change, and seeks to have the smallest carbon footprint possible. It can be easily implemented and is designed to last a hundred years.

But *Creating Green Roadways* is about more than just roads. It is about pedestrian and bicycle facilities, streetscapes, and community character. It is about protecting important cultural and natural resources and ensuring that creatures large and small can cross the road safely. It is about multimodality, natural processes, and energy efficiency.

Why We Need Green Roadways

The automobile is an inherent part of the American way of life. We love our cars. The American Dream includes a white picket fence, a chicken in every pot, and two cars in every garage. For every teenager, obtaining a driver's license is a rite of passage that opens up a whole new world of freedom and possibilities.

So much of our culture has been influenced by the automobile. Roads dominate the American landscape, and we build our cities and towns around the automobile. For the last seventy years or so, it seems

FIGURE 1-1 Using landscape medians, preserving existing vegetation, and planting new trees helps reduce the visual impact of roadways. Image acquired from depositphotos.com.

we have adopted the philosophy that the more roads we have, the better. The development of the Interstate Highway System in the 1950s helped connect the country and resulted in a significant increase in privately owned automobiles. It also led to the creation of the suburb, and the development patterns in the United States changed almost overnight.

FIGURE 1-2 The Blue Ridge Parkway is a 469-mile national parkway that runs mostly along the famous Blue Ridge Mountains chain of the Appalachian Mountains. The parkway is renowned for how well it fits the existing landscape. Image courtesy of Wikipedia.

Highways are also the backbone of the American economy. The Interstate Highway System has served as the nation's primary transportation network for more than half a century, and it has had a lot to do with the diverse, robust economy that we have in the United States.

We rely much more heavily on roads for commercial and personal transit than any other country in the world. Transportation systems provide direct economic and social opportunities and benefits. According to the U.S. Department of Transportation, "Transportation's vital importance to the U.S. economy is underscored by the fact that more than $1 out of every $10 produced in the U.S. gross domestic product is related to transportation activity." Virtually all goods and services involve interstate highways at some point.

Local roads are essential for the day-to-day lives of many Americans. We seem to drive everywhere—to work, to school, to stores, and to parks. As a general

FIGURE 1-3 Montague Center is an attractive and historic New England village. Local roads are designed for slow-moving traffic, and wide shoulders provide for on-street parking. Image courtesy of National Scenic Byways Online (www.byways.org).

rule, when transportation is efficient, the potential market for a given product or service increases, and this leads to a stronger economy. It is important to be able to transport supplies to manufacturers and to ship final products to consumers. The more efficient this system, the better.

The Expansion of Our Highway Infrastructure

When Henry Ford introduced his Model T in 1908, it changed the American culture overnight. It no longer took days to visit relatives in the next town over, and the concept of distance changed significantly. According to the Antique Automobile Club of America, in 1910, Ford produced nineteen thousand Model Ts. By 1927, fifteen million Model T Fords had rolled off the assembly line. For comparison, there were a total of 4,192 automobiles registered in the United States in 1900, and around 210,000 in 1911.

One given is that if you build cars, you need to build roads.

The Antique Automobile Club of America also notes that there were only ten miles of paved roads in the United States in 1900. With all of the new automobiles being produced, there was a demand for more roads. The Federal Highway Act of 1921 provided funding to help state highway agencies construct paved, two-lane interstate highways. During the 1930s, the Bureau of Public Roads helped state and local governments create Depression-era road projects that employed as many workers as possible.

President Franklin D. Roosevelt signed legislation authorizing a network of rural and urban express highways in 1944, but the legislation lacked funding, so nothing was built. When President Eisenhower signed the Interstate Highways and Defense Act in 1956, it was a significant moment in the history of the United States. From the start, the Interstate Highway System was hailed as the greatest public works project in history. The system was constructed as part of the

Figure 1-4 The Tom Moreland Interchange is the intersection of Interstates 85 and 285 in Atlanta. It is also known as Spaghetti Junction, a nickname commonly used in various cities to reference these types of intertwined interchanges because they resemble a plate of spaghetti. Image courtesy of AECOM.

nation's strategic homeland defense, illustrating the important role of transportation in emergency mitigation, defense, and recovery.

Eisenhower's vision of national prosperity was successful beyond expectations. The construction of the Interstate Highway System and subsequent expansion of connected urban freeway systems completely changed how people got around in this country.

A lot has changed since the Interstate Highway System was developed. The U.S. population has almost doubled, increasing from 169 million to more than 300 million, and gross domestic product has increased from $345 billion to $14.3 trillion. In 2006, we traveled more than three trillion vehicle miles, five times the level in 1955 (U.S. House of Representatives, Committee on Transportation and Infrastructure, 2009). In 2009, there were more than four million miles of paved highway in the United States, and the greatest distance from a road in the contiguous states was twenty-two miles (in the southeast corner of Yellowstone National Park in Wyoming). In 97 percent of the continental United States, you were no more than three miles from a paved road.

The Problem

Our current transportation system is broken.

We need a new approach, because the old approach to designing and building roads no longer works. We are not able to meet the transportation needs of today, and the situation is just going to get worse.

The American Society of Civil Engineers estimates that we need to invest $2.2 trillion in federal, state, and local funds over a five-year period to meet transportation requirements. Population growth has created capacity demands that our existing transportation infrastructure has not been able to meet. Funding for transportation infrastructure is woefully inadequate. In early 2009, economist Michael Hudson said, "The [US] economy has reached its debt limit and is entering its insolvency phase. We are not in a cycle but [at] the end of an era."

In the next few years, as our aging infrastructure is repaired and expanded, there will be opportunities to create green roadways. A lot of freeways were built more than fifty years ago and are in a constant state of disrepair. According to TRIP, a nonprofit transportation research group based in Washington, D.C., our transportation infrastructure is badly in need of repair:

- 33 percent of the nation's major roads are in "poor or mediocre condition"
- 36 percent of major urban highways are congested
- 26 percent of bridges are "structurally deficient or functionally obsolete"

One of the big problems is that demands upon our nation's highways have increased significantly over the years. According to the American Association of State Highway and Transportation Officials (AASHTO), there were approximately 65 million cars and trucks in 1955, and today that number has nearly quadrupled to 246 million (Schoen, 2010). In 1970, motor vehicles on U.S. roads traveled around one trillion miles per year; in 2010 this number had increased to more than three trillion miles each year. During that interval, the total number of miles of paved roads increased by only 1.97 percent.

In 2007, the condition of our transportation infrastructure really hit home when an interstate highway bridge in downtown Minneapolis collapsed into the Mississippi River, killing thirteen people and injuring dozens of others. The bridge collapse got the attention of transportation departments across the country, and many realized they needed to focus more on maintaining existing assets.

Additionally, population growth has created capacity demands that our existing transportation infrastructure has not been able to meet. Major cities are becoming more congested, commuting times are increasing, and problems will continue to get worse because of projected growth. By the end of 2050, the world's cities will see their populations expand by 3.1 billion new residents.

How do we build new roads if we can't take care of what we have now? We can no longer afford to design and build roads the way we once did.

Green Roadways and Quality of Life

There is a need to reevaluate how transportation is accommodated in rural, suburban, and urban communities. Roads and highways have such an impact on our communities that we need to start thinking of them in terms of quality of life. This is a basic concept behind developing green roadways.

Roadway projects often are viewed as blights on the landscape because they are the harbingers of noise and traffic, and physically and visually divide the landscape. Green roadways, however, are based upon the idea that a road can be an asset and can add value to a community's quality of life.

A green roadway should protect:

- natural resources by minimizing the impact of the road, reducing our carbon footprint, preserving trees, and managing water resources
- human-made resources by respecting cultural values
- creatures large and small by preserving habitats and ensuring that animals can cross the road safely

FIGURE 1-5 The collapse of the I-35W Mississippi River Bridge in 2007 was a wake-up call about the need to maintain highway infrastructure in this country. Image courtesy of Minnesota Department of Transportation.

- motorists by creating the safest, most functional road possible
- open space and the existing landscape character by minimizing the road footprint and incorporating parks and conservation areas into transportation projects
- the future of our children by being energy efficient and using resources wisely

One of the best ways to build a green transportation infrastructure is to get people out of their cars. By reducing the amount of time people spend in cars, we reduce air pollution and encourage a greater level of physical activity. And if we are using cars less, we will need to build and maintain fewer roads.

Some transportation experts have gone so far as to say we can't maintain the size of our current road system, and what we should do is actually close down 15–20 percent of existing roads. This kind of approach is not unprecedented: the Base Realignment and Closure process has led to the closure of more than 350 military facilities since 1989. The 2005 commission recommended that Congress authorize another Base Realignment and Closure round in 2015, and then every eight years thereafter. It will be interesting to see if we do the same thing with roads.

FIGURE 1-6 At the intersection of Russell, Brooks, and South, Missoula's "Malfunction Junction" is one of the busiest and most congested intersections in the city. Photograph courtesy of www.pedbikeimages.org / Dan Burden.

If we can reduce the number of cars and trucks on the roads, that will go a long way toward a more sustainable future for our kids.

Overview of the Book

Creating Green Roads is about how to integrate roads, bridges, trails, walkways, and other transportation elements in such a way that they become assets, not liabilities. Transportation systems have an impact on everyone. Whether it is driving to work, taking the kids to soccer practice, riding a bus, planning new communities, protecting the environment, or making it easier for kids to walk to school, roadways are a part of our day-to-day lives.

The book is divided into twelve chapters, including this introductory chapter.

Chapter 2 examines existing transportation policies and the impact they have upon how stakeholders make decisions. This includes federal, state, regional, and local policies, and how they work together. In particular, the chapter focuses on the U.S. National Environmental Policy Act of 1970 and its influence on how transportation projects address potential impacts. Although the National Environmental Policy

Act has been around for years, we are just now starting to understand how it can lead to green, sustainable roadways.

Chapter 3 looks at the basic elements that typically make up a transportation project. Understanding the basics provides us with greater opportunities to make changes that lead to greener roadways.

Chapter 4 explores the design and planning process, how to develop a wide range of green alternatives, and the best approaches for ensuring a green roadway that considers both people and place.

Chapter 5 looks at how individuals and organizations can get involved in the planning, design, and implementation of green roadway projects.

Chapter 6 examines green roadways in urban areas and presents case studies that indicate successful approaches for implementing urban solutions.

Chapter 7 looks at suburban and rural areas and the unique transportation issues they face. This chapter also presents case studies on successful green roadway projects in these settings.

Chapter 8 explores how green roadway projects can be developed while respecting cultural, historic, and visual resources. It emphasizes how to develop green roadways that fit communities and help create a sense of place.

Chapter 9 focuses on how to develop green roadways that are environmentally sensitive and respectful of little creatures, with the aim of actually improving the world around us. This approach means looking at alternative energy sources and thinking of roadways as laying lightly on the land, and perhaps even being carbon-positive.

Chapter 10 looks at how we can construct green roadways so they last longer, are easier to build, and are less disruptive of people and the environment.

Chapter 11 examines the economics of green roadways and what we need to do to build the next generation of green roadways.

Chapter 12 looks at the next steps in creating green roadways. How do we plan for the next generation of green roadways, and what do we have to do to make this work?

Chapter 2

Transportation Policies

Creating green roadways is a cooperative process designed to foster involvement by all users of the system. The most sustainable approach to transportation would be to maximize use of existing public infrastructure, reduce people's need to drive, increase roadway connectivity, disperse traffic, and minimize the construction of new roads.

Traditionally, the role of the federal government in transportation has been to set national policy, provide financial aid, supply technical assistance and training, and conduct research. The federal government provides a significant amount of funding for states to implement transportation projects, and there are requirements attached to those funds. Because of this combination of policies, regulations, and funding, federal agencies have a strong influence on state and local transportation decisions.

Green roadway projects are conducted at state and local levels. This is entirely appropriate because highway and transit facilities and services are owned and operated largely by the states and local agencies.

Federal Policies and Procedures

The federal transportation policy framework consists of a series of laws, regulations, orders, and other documents, and a number of different agencies are involved with implementing and enforcing these. It is important to know: (1) the key agencies and organizations that are involved with federal policies and procedures, and (2) the major federal acts and regulations that define transportation policy.

Key Players at the Federal Level

To create a successful green roadway project, it is important to understand the key agencies and organizations at the federal level.

Federal Highway Administration (FHWA)–responsible for administering the federal highway aid program to individual states and working to help plan, develop, and coordinate construction of federally funded highway projects.

Transportation Research Board–a division of the National Research Council whose mission is to promote innovation and progress in transportation through research.

U.S. Department of Transportation–oversees federal highway, air, railroad, maritime, and other transportation administration functions; it consists of ten agencies, including the FHWA and the Federal Transit Administration.

Federal Transit Administration–an agency within the U.S. Department of Transportation that provides financial and technical assistance to local public transit systems.

National Surface Transportation Policy and Revenue Study Commission–charged by Congress under Safe, Accountable, Flexible, Efficient Transportation Equity Act: A Legacy for Users (SAFETEA-LU) to determine future needs of the surface transportation system.

National Surface Transportation and Infrastructure Financing Commission–created by Congress to study highway and transit needs and make recommendations for future actions.

American Association of State Highway and Transportation Officials–sets technical standards for all phases of highway system development.

Council on Environmental Quality–created by Congress to oversee the federal implementation of the National Environmental Policy Act (NEPA).

U.S. Environmental Protection Agency (USEPA)–is in charge of restoring, protecting, and enhancing the environment to ensure public health, environmental quality, and economic vitality in the United States.

U.S. Institute for Environmental Conflict Resolution–established by the Environmental Policy and Conflict Resolution Act of 1998 to assist in resolving conflict over environmental issues that involve federal agencies.

U.S. Army Corps of Engineers–is responsible for investigating, developing, and maintaining the nation's water and related environmental resources.

National Cooperative Highway Research Program–conducts research in problem areas that affect highway planning, design, construction, and maintenance.

American Public Transportation Association–a group of public organizations engaged in busing, transit, light-rail, commuter rail, subways, waterborne passenger services, and high-speed rail.

Intelligent Transportation Systems Institute–established under the Intermodal Surface Transportation Efficiency Act (ISTEA), this institute is housed at the University of Minnesota and focuses on human-centered transportation technology, such as computing, sensing, communications, and control systems.

There are dozens of other organizations and agencies involved with major transportation projects because of their importance and complexity. Now add in (1) local, state, and regional transportation agencies; (2) other organizations and agencies involved with related issues such as environmental resources, wildlife, trails, parks and recreation, land use planning, utility infrastructure, economics, and regulatory requirements; and (3) stakeholder groups interested in specific projects. Trying to get this many groups with this many different interests to agree is difficult, but it can be done, as shown by the case studies in this book.

Federal Acts and Regulations

As mentioned earlier, one of the major roles of the federal government in transportation planning is to set national policy. The country's transportation policy has been defined by a series of acts and regulations that determine how, where, and when we develop transportation projects. These include the following:

Interstate Commerce Act of 1887–regulates the railroad industry.

Federal-Aid Highway Act–a set of bills and acts of Congress that have funded transportation in the United States; the first was offered in 1916.

Federal-Aid Highway Act of 1921 (Phipps Act)–defined the Federal-Aid Road Program to develop an immense national highway system.

National Interstate and Defense Highways Act (1956)–authorized the building of highways throughout the nation and established the Interstate Highway System.

Federal Highway Administration. Code of Federal Regulations Title 23—Highways, Part 752 Landscape Development (1960)–establishes landscape development policy and guidelines on federally funded or federally participating construction projects.

Federal-Aid Highway Act (1962)–created the federal mandate for urban transportation planning in the United States.

Urban Mass Transportation Act of 1964 and 1970–provided funding for urban transit.

Highway Beautification Act of 1965–called for control of outdoor advertising and encouraged scenic enhancement along roads.

National Historic Preservation Act (1966)–the primary federal law governing the preservation of cultural and historic resources in the United States.

Architectural Barriers Act of 1968 and the Americans with Disabilities Act of 1990–establish guidelines for accessibility of facilities.

U.S. National Environmental Policy Act (NEPA) of 1970–establishes the broad national framework for protecting our environment.

Clean Air Act of 1970–primary legislation that controls reduction of airborne contaminants—smog and air pollution in general.

Clean Water Act of 1972–the primary U.S. federal law governing water pollution.

Noise Pollution and Abatement Act of 1972–regulates noise pollution with the intent of protecting human health.

Endangered Species Act (1973)–provides a program for the conservation of threatened and endangered plants and animals and the habitats in which they are found.

Surface Transportation Assistance Act of 1982–extended authorizations for the highway, safety, and transit programs and governed the movement of trucks and trailers with specific combinations, lengths, or widths.

Surface Transportation and Uniform Relocation Assistance Act of 1987–the last authorization bill of the interstate era.

Farmland Protection Act (1988)–minimizes the extent to which federal programs contribute to the unnecessary conversion of farmland to nonagricultural uses.

Intermodal Surface Transportation Efficiency Act (ISTEA)–introduced in 1991, established criteria for developing transportation plans and programs.

Transportation Enhancement Program (1992)–promotes trails, landscaping, and historic preservation.

Environmental Justice Executive Order (1994)–directs federal agencies to make achieving environmental justice part of their mission.

National Highway Designation Act (1995)–Congress approved the National Highway Designation, which included 160,955 miles of roads, including the Interstate Highway System, as the National Highway System.

Transportation Equity Act for the 21st Century (TEA-21) (1998)–successor legislation to ISTEA.

Safe, Accountable, Flexible, Efficient Transportation Equity Act: A Legacy for Users (SAFETEA-LU)–enacted on August 10, 2005, it authorized $286.4 billion for transportation projects for the five-year period 2005–2009.

Hiring Incentives to Restore Employment Act–jobs bill that extended Highway Trust Fund through December 31, 2010.

When it comes to creating green roadways, some acts and regulations are often an integral part of design and planning discussions. More detailed information about some of these acts and regulations follows.

ISTEA

ISTEA, which was introduced in 1991, established new criteria for developing transportation plans and programs. ISTEA divided roads into two categories: those that are part of the National Highway System (NHS) and those that are not.

The NHS consists of major streets and roads around the country—interstate highways and other routes with heavy motor vehicle traffic. Under ISTEA, working on routes that are not part of the NHS—meaning most streets and roads—is no longer subject to any federal design rules. Such work can be funded with federal money but is subject only to state design rules.

According to the National Transportation Library, among other provisions, ISTEA

- established the National Highway System to focus federal resources on roads that are the most important to connectivity
- ensured more flexibility for state and local governments in determining transportation, planning, and management systems
- ensured that funding would be available for bicycle and pedestrian transportation as well as other enhancements
- promoted new technologies such as intelligent vehicle highway systems
- worked with the private sector as a funding source
- provided discretionary and formula funds for mass transit
- provided highway funds to enhance the environment

Transportation Enhancement Program

Since the 1992 inception of the Transportation Enhancement (TE) program (part of ISTEA) through 2009, approximately $9.2 billion has been spent on more than 24,000 projects throughout the country. The priorities for the TE program are (1) bicycle and pedestrian facilities, (2) landscaping and beautification, and (3) historic preservation and rehabilitation. In fact, hike- and bike-oriented projects accounted for 56.4 percent of program funding between 1992 and 2009. These are the types of site amenities that help make green roadways green.

Like other components of the Federal-Aid Highway Program, TE activities are federally funded and state administered. According to the governing legislation, there seem to be two points of consensus when it comes to TE funding: (1) It is money well spent. Almost everyone seems to support the ideas of trails, landscaping, and improving visual quality. (2) The amount of funding for transportation enhancements is inadequate.

TEA-21

The Transportation Equity Act for the 21st Century, commonly referred to as TEA-21, was the successor legislation to ISTEA. It continued most of ISTEA's programs and policies and maintained an emphasis on local involvement in transportation decision making. TEA-21 authorized $218 billion in federal funding for the period between October 1, 1997 and September 30, 2003. Although TEA-21 was to end in 2003, the act was extended through 2005.

TEA-21 used the term "context-sensitive design" in reference to the fundamental idea that every highway project is unique and that highway development should be integrated with communities and the environment while maintaining safety and performance. For transportation engineers, this approach was nothing short of radical at the time. For the next generation of green roadways, context is an essential ingredient in the process.

FIGURE 2-1 The New River Gorge Bridge in Fayetteville, West Virginia, was completed in 1978. One of the largest arch bridges in the world, it is an icon and symbol of pride to the people of West Virginia. Photograph by Steve Shaluta, courtesy of National Scenic Byways Online (www.byways.org).

TEA-21 also established the Transportation and Community and System Preservation Pilot Program. Fundamentally, the program is intended to help communities address transportation needs, considering community character, land use, environmental quality, and economic development, and combine all of these issues into a holistic, integrated plan. A total of $120 million was authorized for this program for fiscal years 1999–2003, and FHWA is considering adding additional discretionary funds.

SAFETEA-LU

SAFETEA-LU (2005) authorized $286.4 billion for highways, highway safety, and transit for the five-year period 2005–2009.

SAFETEA-LU placed a greater importance on addressing environmental issues during the transportation planning process; introduced the concept of "participating agencies" to get governmental agencies more involved in the NEPA process; created the Safe Routes to School Program; required each state to establish and implement a schedule of highway safety improvement projects and strategies for hazard correction and prevention; and included new regulations on historic or parkland properties (known as Section 4[f])).

NEPA

NEPA is one of the most important pieces of legislation of this century. It has a significant impact on every project in the United States that requires federal funding or permitting, including transportation projects. NEPA initiated a national policy to "create and maintain conditions under which [humans] and nature can exist in productive harmony, and fulfill the social, economic, and other requirements of present and future generations of Americans."

For green roadway projects, it is important to understand what NEPA requires, allows, and prohibits. NEPA is intended to promote roads that are more environmentally, culturally, and visually sensitive to their surroundings. The key is to know how to use NEPA to make sure that happens.

NEPA intends that impacts to the natural and human environment should be viewed and assessed through a multidisciplinary approach. NEPA mandates that federal agencies (1) consider the potential environmental consequences of their proposals, (2) document the analysis, and (3) make this information publicly available. The NEPA process must be completed before an agency makes a final decision on a proposed action. NEPA procedures are available online at the NEPAnet website (http://ceq.hss.doe.gov/nepa/nepanet.htm/).

Three federal agencies implement NEPA. The Council on Environmental Quality oversees the federal implementation of NEPA, principally through issuance and interpretation of NEPA regulations. The council establishes minimum requirements and calls for agencies to create their own implementing procedures. The USEPA's Office of Federal Activities reviews environmental impact statements (EIS) and some environmental assessments (EA) issued by federal agencies. Comments received by the USEPA are

FIGURE 2-2 Red and gold leaves combined provide a rich color amongst the fall foliage. Photograph by Gary Johnson, courtesy of National Scenic Byways Online (www.byways.org).

published in the *Federal Register*, a daily publication that provides notice of federal agency actions. Third, the U.S. Institute for Environmental Conflict Resolution, established in 1998, assists in resolving conflicts over environmental issues that involve federal agencies.

Environmental Impacts

The potential impacts of a major transportation project include the destruction of natural resources, an increase in stormwater runoff and resulting decrease in water quality, an increase in noise and air pollution, fragmentation of wildlife habitat, and the disruption of natural processes. The impacts that NEPA requires be addressed fall into three main categories:

Direct impacts–Those "which are caused by the action and occur at the same time and place." An example is damage to a wetland due to the reconstruction of a bridge.

Indirect and cumulative impacts–The degree to which indirect and cumulative impacts need to be addressed in a NEPA document depends on the potential for the impacts to be significant and will vary by resource, project type, geographic location, and other factors. An example of indirect effects of a new interchange project could include potential changes in land use such as a new gas station that would not have been built without the interchange.

Cumulative impacts–Those that would result from the incremental impact of the project when added to other past, present, and reasonably foreseeable future actions. This could include increased air and water pollution that result from a new highway.

According to the Tennessee Department of Transportation, impacts are defined differently for each subject. For example, noise impacts are classified as "substantial," "moderate," or "minor," and visual impacts are "high," "medium," and "low." Historic resource impacts are classified as "no historic properties affected," "no adverse effect," or "adverse effect."

When negative impacts are identified during the planning process, there are three basic approaches for addressing them: (1) avoidance, (2) minimization, and (3) mitigation.

Avoidance–The best approach for any green roadway project is to avoid affecting sensitive cultural, natural, and visual resources. Road alignments can be shifted, or the footprint of a road project can be minimized by reducing lanes, using retaining walls, adjusting slopes, or implementing other elements.

Minimization–There are times when it is not possible to totally avoid impact to sensitive resources, but measures can be taken to keep these impacts to a minimum. Green roadways can be designed in such a way that they "fit" their surroundings while causing minimal impacts. The case studies in this book illustrate dozens of ways minimization can be handled.

Mitigation–It is not always possible to avoid or minimize impacts. In these situations, emphasis is on mitigation by way of compensation and enhancement. If a road affects an existing wetland, one mitigation approach is to build a constructed wetland somewhere else to compensate for the loss. If a vegetated area is affected, it can be restored by using native plant materials planted using ecological principles.

Classes of Action

For transportation projects that are supported by federal funds, NEPA requires that all reasonable alternatives be identified and analyzed. This is an important distinction, because it refers only to alternatives that seem logical and can be implemented. Regardless of size, complexity, or type of transportation project, environmental impacts are addressed in NEPA in one of three actions: environmental impact statement, environmental assessment, or categorical exclusion.

Figure 2-3 This bridge in New Mexico uses native stones, a low horizontal design, and subdued brown and tan colors to help visually integrate into the surrounding landscape. Image courtesy of EDAW.

Environmental Impact Statement

An EIS must be prepared for every major federal action that may significantly affect the quality of the human environment. An EIS is often required because of anticipated impacts on sensitive environmental areas such as wetlands, valuable habitat for wildlife, old growth forests, or culturally or historically important sites.

One common misconception is that an EIS is required for most transportation projects, and they significantly slow down the planning process and the development of new roads and trails. According to the FHWA, the reality is that only about 3 percent of federally funded transportation projects require an EIS. These are the projects in which significant environmental impacts are anticipated. An EIS typically takes two to three years to complete, with much of that time used for mandatory reviews.

A supplemental EIS (SEIS) is necessary when major changes, new information, or further developments occur in the project that would result in significant environmental impacts not identified in the most recently distributed EIS. Three triggers necessitate an SEIS: the project is proceeding to the next major federal approval, the project changes, or the three-year timeline for an EIS is nearly over and additional study is needed.

For both an EIS and SEIS, the process usually involves a draft and a final document. The draft EIS is developed and submitted for review. Once the public comment period is finished, the lead agency for a specific project analyzes comments, conducts further analysis as necessary, makes appropriate changes, and prepares a final EIS.

Environmental Assessment

An EA is prepared for projects that may have adverse effects on the environment that are not as significant as the potential effects of a project requiring an EIS. For example, for a project requiring an EA instead of an EIS, only the perimeter of a wildlife habitat area may be affected, and no rare and endangered species are present. An EA is intended to be a concise document that (1) briefly provides sufficient evidence and analysis for determining whether to prepare an EIS, (2) aids an agency's compliance with NEPA when no EIS is necessary, and (3) facilitates preparation of an EIS when one is necessary.

An EA process typically takes six to eighteen months to complete and concludes with either a Finding of No Significant Impact or a determination to proceed to preparation of an EIS.

Categorical Exclusion

A categorical exclusion is a category of actions that does not individually or cumulatively have a significant enough effect to warrant action via NEPA. A categorical exclusion is determined only if there is no question about potential impacts. If the lead agency is unsure about potential impacts, then an EA should be conducted. Approximately 90 percent of all surface transportation projects have minimal environmental impacts and therefore receive categorical exclusions.

NEPA Process for Green Roadways

Although NEPA has been around for more than forty years, it seems that we just now understand how to use it as part of a more holistic, integrated planning approach for transportation projects. The first twenty years of NEPA were spent trying to understand how the process worked.

NEPA is especially important for major transportation projects because they typically are complex, cover a large geographic area, and can have a major impact on cultural, natural, and visual resources. NEPA provides the legal foundation to ensure that transportation projects take those resources into account. When it comes to creating green roadways, NEPA can be used to go beyond the status quo. One of the primary ways that NEPA allows this is by providing a transparent, inclusive process that gives key stakeholders an opportunity to participate. A series of "concurrence points" defines milestones at which all federal and state agencies involved in a project must reach agreement before the next phase of the NEPA process can begin. This means that agencies concerned about historic resources or aesthetics, for example, also have a voice in determining the potential impacts and benefits of a project.

Following this process doesn't automatically guarantee that the resulting transportation project will be one that is green, but it does provide the framework that makes such a project possible. The best thing about the NEPA process is that there aren't any surprises. The Concurrence Points require that agencies actually sign a letter of agreement to minimize potential misunderstandings and make sure that each agency was involved.

The most successful transportation projects are those in which agencies and stakeholders are most involved and in which the planning team includes a wide range of disciplines as part of an integrated team. One major benefit of having such a diverse group at the same table is that the alternatives generated are usually more creative and explore a wider range of possibilities. Starting off with the philosophy that there are no bad ideas helps ensure that concepts are not dismissed too quickly without being given due consideration.

Sections 4(f) and 6(f) Requirements

In addition to general impact assessments, certain types of land are singled out as needing special consideration. NEPA Section 4(f) protects significant

Box 2.1 NEPA Process

The NEPA process is as follows:

1. Define Purpose and Need–A purpose and need statement is the foundation of the NEPA decision-making process. It has three parts: the purpose, the need, and the goals and objectives. The purpose defines the transportation problem to be solved. The need establishes evidence that a problem exists. The goals and objectives identify desirable outcomes.
2. Concurrence Point 1–Participating agencies provide a response to purpose and need. Modifications are made until all agencies agree to sign off on the document so the NEPA process can continue.
3. Scoping–The scoping process is intended to determine the scope of issues to be addressed for a proposed action.
4. Identify Alternatives–During this phase, the basic options to be considered are identified.
5. Concurrence Point 2–Participating agencies review alternatives (within forty-five days) and respond.
6. Analyze Reasonable Alternatives–The alternatives and potential impacts are analyzed to determine if an EIS, an EA, or a categorical exclusion is required.
7. Prepare a Draft EIS
8. Includes all information developed through the NEPA process up to this point.
9. Concurrence Point 3–Participating agencies review the draft EIS (within forty five days) and respond.
10. Hold Public Hearings on Draft EIS–Draft EIS is shared with the public via a series of meetings to give stakeholders an opportunity to participate in the planning process.
11. Determine Preferred Alternative and Mitigation Measures–Alternatives are analyzed to determine which one best meets the purpose and need for the project while minimizing impacts on cultural, natural, and visual resources.
12. Concurrence Point 4–Participating agencies review and provide concurrence on Preferred Alternative and Preliminary Mitigation (within forty five days).
13. Prepare Final Environmental Document–State Department of Transportation prepares the Finding of No Significant Impact or final EIS for appropriate approvals.
14. Submit Record of Decision (ROD)–The final step in the EIS process is for the agencies involved to determine a record of decision.

historic sites as well as publicly owned parks, recreation areas, and wildlife and waterfowl refuges from being affected by transportation projects or programs, unless (1) there is "no prudent and feasible alternative" to harming the site, and (2) the project includes "all possible planning to minimize harm" to the protected resources. Coordination with local officials and agencies is a critical part of the Section 4(f) evaluation process. An open dialogue can help identify potential conflicts and help lead to possible resolution. Section 4(f) compliance has historically been one of the most litigated components of the transportation planning process. Because of the litigious nature of Section 4(f), it is important that all decisions and discussions be clearly documented.

The FHWA is ultimately responsible for making all decisions related to Section 4(f) compliance. Before approving a project that uses Section 4(f) property, the FHWA must either (1) determine that the impacts are "de minimis," meaning they don't have an adverse impact, or (2) undertake a Section 4(f) evaluation. Once a project's impacts are deemed to be de minimis, further evaluation is not required.

Section 6(f)

Section 6(f) of the Land and Water Conservation Fund Act of 1965 is intended to restrict the future use of parklands or open spaces. Section 6(f) is administered by the National Park Service. A state Department of Transportation (DOT) is required to work with the park service on any project that may affect publicly owned parkland, which is considered to be any land acquired and/or developed with Land and Water Conservation Fund monies.

As a general rule, Section 6(f) is not as stringent as 4(f) and is meant to ensure that there is no overall loss in parkland and open space.

NEPA and Public Involvement

Every major step in the NEPA process includes an opportunity for public input, in recognition that public involvement is a critical part of every planning project. NEPA stresses that for public involvement to be effective, a meaningful open exchange of information and ideas between the public and transportation decision makers needs to happen. The consensus is that public involvement and more stringent environmental reviews have improved the quality of roadway projects. (See chapter 5 for more information on stakeholders and partnerships.)

One of the best times for interested stakeholders to get involved in a transportation project is during the NEPA process. For example, public comments on draft EAs and EISs provide a basis for refinement of proposed mitigation strategies in NEPA processes.

In general, it is easy for the public to monitor the development process of major transportation projects. Every major project usually has a website that includes all of the information required by NEPA as well as a schedule of future meetings, and information is often printed in local newspapers and in the *Federal Register*.

Environmental Streamlining

The two biggest complaints about the NEPA process are that it takes too long and costs too much. The concept of "environmental streamlining" is based on the idea that environmental issues can be addressed holistically during the transportation planning process in a more timely and cost-effective way. Environmental streamlining strives to improve project delivery without compromising environmental protection.

It typically takes years for an EIS to be completed and approved because of the scale and complexity of many transportation projects. According to the FHWA, environmental reviews have taken an average of about six years since fiscal year 1999, and about 12 percent have taken longer than ten years. The FHWA conducted research in 2007 and discovered that it takes two to three years to prepare an EIS, but the average time taken to complete the statements, conduct reviews, and gain final approval was about seven years.

Several efforts have been implemented to make these reviews more efficient and timely without diminishing environmental protections, and these have been very effective.

FHWA's Vital Few Environmental Streamlining and Stewardship goal is focused on improving processes that influence outcomes and benchmark the results of significant stewardship activities. The FHWA has the final approval and determines when an EIS is in compliance with applicable environmental laws and other requirements.

FHWA and the Federal Interagency Workgroup developed the Environmental Streamlining National Action Plan (2006), which includes five elements:

- program efficiency
- mitigation strategies
- resource management
- continuous improvement
- dispute resolution

The National Memorandum of Understanding for Environmental Streamlining was written and signed to implement Section 1309 of TEA-21. Under the memorandum of understanding, federal resource agencies such as the U.S. Fish and Wildlife Service, U.S. Department of Agriculture, U.S. Department of Commerce, USEPA, U.S. Army Corps of Engineers, and the Advisory Council on Historic Preservation all agreed to work together to conduct concurrent project reviews under NEPA and other legal authorities' approvals.

Planning and Environmental Linkage is defined by the FHWA (2008) as a voluntary approach to transportation decision making that considers environmental, community, and economic goals early in the planning stages and carries them through project development, design, and construction. The planning and environmental linkage process incorporates a process similar to the one required by NEPA, but it does not require a project funding commitment.

NEPA/404 Merger Process is designed to consolidate the NEPA environmental review process with permitting requirements of Section 404 of the Clean Water Act. This will help streamline the amount of time it takes to complete the review process. One of the most commonly cited benefits of the merger process is increased communication and understanding of the issues involved. The merger process requires concurrence of all parties involved, and this can be an issue because there are differences in the purpose and need as defined by NEPA and that defined by Section 404.

Future of NEPA

Although the number of successful green roadways that have used NEPA in a variety of ways to improve performance and protect important resources is growing, there are some concerns about the future impact of the law. Since 2008, the federal budget for transportation projects has been axed, and reduction of available transportation funds may have an impact on how environmental resources are protected. As detailed in this chapter, the standard NEPA process requires a comprehensive environmental study to be conducted on every transportation project unless it is classified as a categorical exclusion. Because this is a time-consuming and expensive process, it can cost millions of dollars to conduct a full-blown EA. There is concern by some that transportation projects may be put on hold because of the high cost of meeting NEPA requirements, or that there may be requests to "temporarily" waive NEPA requirements in order to save money.

Another concern is that too often the process taken to meet NEPA requirements is not integrated with other planning efforts, such as land use plans, parks and recreation master plans, water management plans, and long-range transportation plans.

NEPA guidelines have resulted in roads that function more efficiently and avoid harmful environmental impacts. Despite cost-cutting pressures, not taking the time to examine these impacts and plan around them will be more costly in the long run because of environmental issues.

It is important that NEPA not only remain effective legislation, but that it be applied more efficiently to create greener, more sustainable projects.

A good example of how NEPA can help create a better transportation project is the planned expansion of Michigan's US Route 23 from a two-lane to a four-lane freeway, which met resistance from the very start. The case study comes from the Sierra Club Foundation's report *The Road to Better Transportation Projects: Public Involvement and the NEPA Process*. The proposed expansion was to have affected undeveloped land, compromised important wildlife habitat, and resulted in the largest single wetlands loss the state had ever experienced. The Michigan Department of Transportation presented the only two viable options as building the freeway expansion or doing nothing. The FHWA ruled that the Michigan Department of Transportation failed to adequately look at other alternatives as required by NEPA. The FHWA recommended a more context-sensitive approach that included passing lanes, traffic signal upgrades, and turn lanes to improve the road. This approach saved $1.5 billion and resulted in a road more in keeping with the desires of local stakeholders.

State and Local Policies and Procedures

State transportation policies are developed within a framework of federal and state statutes, laws and regulations, and plans. Each state has a transportation department that is responsible for planning, programming, implementation, and maintenance, and works cooperatively with tolling authorities, ports, local agencies, special districts, and others involved with transportation.

Each state DOT establishes its own guidelines and requirements for providing the nonfederal share of project costs. Some states require local sponsors to provide a share of project costs. Most states' reports have a long list of transportation enhancement projects that they have funded. Each state is able to prioritize its projects to determine which ones to implement first.

Primary transportation planning functions of the state DOT include preparing and maintaining a long-range statewide transportation plan, developing a Statewide Transportation Improvement Program, and involving the public.

State Programs

The Surface Transportation Program first established as part of ISTEA is a funding mechanism that can be used statewide for qualified transit projects.

A Strategic Highway Safety Plan is a statewide safety plan that identifies the key safety needs and guides investment decisions to achieve significant reductions in highway fatalities and serious injuries on all public roads. One major benefit of such a plan is that it leads to collaboration among transportation planners, traffic engineers, landscape architects, environmentalists, and stakeholders.

Metropolitan Planning Organizations

A metropolitan planning organization (MPO) is a federally mandated and federally funded transportation policy-making organization consisting of representatives from local governments and transportation authorities. Federal legislation in the early 1970s required the formation of an MPO for any urbanized area with a population greater than 50,000.

In metropolitan areas, the MPO is responsible for actively seeking the participation of all relevant agencies and stakeholders in the planning process; the state DOT is responsible for these activities outside metropolitan areas.

There are five core functions of an MPO:

- Establish and manage a fair and impartial setting for effective decision making.
- Identify and evaluate alternative transportation improvement options.
- Prepare and maintain a Metropolitan Transportation Plan covering at least twenty years.
- Develop a short-range Transportation Improvement Program based upon the Metropolitan Transportation Plan.
- Involve the public in the transportation planning process.

Some MPOs are found within agencies such as regional planning organizations, councils of governments, and others. Areas with populations greater than 200,000 are designated transportation management areas, and they must also include a congestion management process as part of their responsibilities.

State/Tribal Historic Preservation Officers

Staff of the federal agency responsible for a project must determine whether an activity could affect historic properties. These are properties that are included in the National Register of Historic Places or that meet the criteria for the National Register. If historic properties could be affected, the federal agency must consult with the appropriate state or tribal historic preservation officer during the process. Both agencies have a major impact on protecting cultural and historic features during the transportation planning process.

These transportation policies lay the foundation for decisions about how to design, build, and maintain transportation projects. They also greatly influence our ability to develop green roadway projects.

Chapter 3

Basic Roadway Design

In the United States, we have been designing and building roadways for a very long time, and over the years our transportation engineers' understanding of road geometry has become quite sophisticated.

Geometric design deals with features of location, alignment, profile, cross section, and intersections for a range of highway types and classifications. Rural roads typically have fewer issues than urban roads. The geometric form and dimensions of the highway should properly reflect driver safety, desires, expectations, comfort, and convenience, and do so within the context of the surrounding landscape.

State and federal agencies continuously fine-tune road geometry, and although the basic standards are pretty well set, there are still a lot of design options for creating green roadways.

This chapter starts by looking at broader issues that affect basic roadway design—road types, safety concerns, right-of-way requirements, etc.—then focuses on the components that make up a green roadway.

Road Types

Roadways can be classified as one of the following:

Freeways/interstates–Freeway design standards for interstate highways are established by the Federal Highway Administration (FHWA) because these roads are components of the National Highway System. They are the only class of road on which pedestrian and bicycle travel is prohibited. Entrance and exit ramps are the primary means of accessing freeways and other controlled-access facilities. There is not a lot of flexibility regarding the road geometry of interstate highways, but there are a lot of options for making the highway corridor greener and more attractive.

Arterials and major collectors–Arterials and major collectors vary widely in character depending upon where they are constructed. Travel lanes of eleven or twelve feet are usually provided on arterials and major collectors.

Minor collectors–Minor collectors are often designed for low-speed, low-volume operations. They are often designed to accommodate traffic between major collectors, arterials, and freeways.

Local roads–Local roads are generally constructed to comply with municipal requirements. However, the guidance provided for arterials and major collectors is suitable for local roads with high volumes and high speeds. Much like minor collectors, local roads are

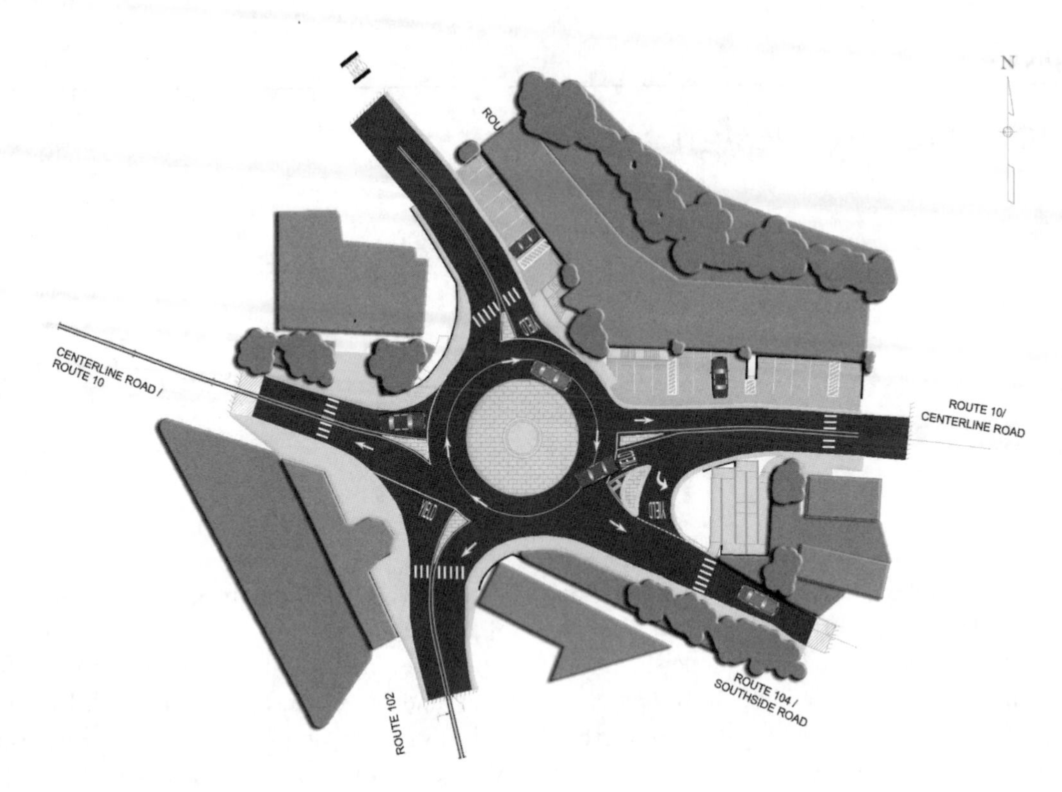

FIGURE 3-1 This is an example of a typical interstate highway exchange with on- and off-ramps and an overpass spanning the highway. Image acquired from depositphotos.com.

sometimes designed to provide shared accommodation for all users.

Parkways–Parkways are typically located within parks or parklike settings that place an emphasis on natural resources, visual quality, and recreational opportunities. Parkways often have wider rights-of-way than other roads; this provides opportunities for preserving existing natural resources as well as adding landscaping and other amenities. Frequently, parkways are part of linear parks that include trails, forested areas, and open meadows.

Green roadways–The concept of a green roadway can be applied to any type of road. Green roadways are as much a philosophy and an approach as anything else. It is possible to build all types of roads so that they protect and enhance the environment.

One major focus of green roadways should be to make communities more livable. FHWA defines "livable communities" as "places where the young and old can walk, bike, and play together; where historic neighborhoods are preserved; where farms, forests, and other green spaces are protected; where parents spend less time in traffic and more time with their children, spouses, and neighbors; where older neighborhoods can thrive once again. A livable community has safe streets, good schools, and public and private spaces that help foster a spirit of community."

Access Control

Access control defines the degree to which properties are connected to a roadway. The more we can control how vehicles move within both an urban and rural

FIGURE 3-2 Looking along US Route 44 south of Manila, Utah, the road follows the base of the mountains to minimize cutting into the slope and stay out of the adjacent river valley. Image courtesy of National Scenic Byways Online (www.byways.org).

settings, the more we can ensure that a roadway fits within a community. Controlling access limits the overall impact cars have and provides more options for trails, walkways, and public spaces. Access management applies roadway and land use techniques to preserve the safety, function, and capacity of transportation corridors. The objective is to ensure roadway safety and efficient operations while providing reasonable access to adjacent land uses.

Massachusetts Department of Transportation cites the following degrees of access control:

Controlled access–Freeways or other major arterials where access to the roadway is limited to interchange points or major intersections.

Limited access–Typically arterials where intersections are widely spaced and driveway connections are limited. Driveways to properties may be consolidated to limit connections to the roadway. Major intersecting streets may be signalized or handled at interchanges (note that at an interchange, the traffic does not stop moving). Minor intersecting streets may be limited to right-turn-only or may be grade-separated.

Full access–Typically arterials or collectors where access is provided to adjoining properties without restrictions on turning movements. Driveway spacing and other design guidelines are typically applied. Intersecting streets usually provide the full complement of turning movements.

Uncontrolled–Typically collectors and local roads where access controls are not employed.

Access management plans not only help create a safer road, they create a greater sense of organization

FIGURE 3-3 This cobblestone street in Lyngby, Denmark, provides the city with a beauty and charm that attracts shoppers and creates a timeless experience. Image courtesy of www.pedbikeimages.org / Dan Burden.

and consistency. The Transportation Research Board's *Access Management Manual* (2003) provides guidance on the application of access management techniques.

Types of Transportation Projects

Regardless of the type of road, there are four basic types of transportation improvement projects.

New construction–New construction projects involve building from scratch. Forty years ago, most of the major transportation projects involved new construction, but that trend has shifted and we are spending more time reconstructing existing roads instead of building new ones. New construction often involves a bypass, toll way, or additional road to improve service or accessibility. One major problem with new construction is that it is typically located in undeveloped areas, and the environmental and cultural impacts could be significant.

Reconstruction–Reconstruction projects typically involve significant changes to an existing highway. Often reconstruction is required to widen a highway by adding additional travel lanes to meet traffic demands.

Many of the nation's roads are more than fifty years old, and some need to be replaced. A challenge with reconstruction projects is trying to accommodate existing traffic needs during the reconstruction project.

Resurfacing, restoration, rehabilitation (3R)–These types of projects focus on improving existing transportation facilities in an effort to extend the service life of those facilities. 3R projects disrupt existing traffic, but only for short periods of time.

Maintenance–Once a project is constructed, regular maintenance needs to occur to keep a road operational. Although 3R projects address major issues, there are also smaller issues that are still important in maintaining a safe, accessible road. These activities include mowing, litter pickup, watering, landscape installation and maintenance, salting in winter months, inspections, and day-to-day maintenance of structures such as bridges, retaining walls, signage, guardrails, or other facilities.

Sources of Basic Transportation Standards

There is no shortage of information about geometric road design, but the majority of the publications available are very technical in detail and are geared toward transportation engineers. One problem is that in the past, transportation standards were often interpreted very rigidly, resulting in roadways that were too utilitarian.

The American Association of State Highway and Transportation Officials' (AASHTO) *Policy on Geometric Design for Highways and Streets,* typically referred to as the "Green Book" (yes, it has a green cover), is the foundation for highway design in the United States. The Green Book has been adopted by FHWA as the national standard for roads on the National Highway System. Virtually all state highway design manuals are either derived from or directly reference the AASHTO Green Book. The benefit of this approach is that there is a level of consistency in how roads are designed, including road width, curve radius, design speeds, and other design features.

AASHTO has emphasized that the Green Book is intended to be a flexible document, but unfortunately, many state design manuals based on this book have become more rigid over the years. One reason is that states take a very conservative approach to limit liability and keep costs down. To many, increased flexibility is synonymous with increased risk, but research has clearly shown that is not the case. Creating green roadways requires that flexibility be considered in order to create a better project.

The AASHTO *Roadside Design Guide* addresses design of slopes, clear zones and recovery areas, traffic barriers (guardrail, bridge rail, median barrier), roadside hardware, curbs, and median treatments. Another AASHTO publication, *A Guide to Achieving Flexibility in Highway Design,* focuses on the flexibility inherent with AASHTO's Green Book. A key concept of the book is that one size does not fit all when it comes to transportation projects.

A Guide to Achieving Flexibility in Highway Design emphasizes integrating highway improvements with the surrounding human and natural environments. The publication illustrates how to make highway improvements while preserving and enhancing the adjacent land or community. In particular, it offers suggestions on how to go beyond the guidelines in the AASHTO Green Book. *A Guide to Achieving Flexibility in Highway Design* is a good starting point for anyone interested in greener, more sustainable transportation projects. AASHTO has also published a number of other books about roadway design.

However, AASHTO is not the only source of geometric design criteria. The Institute of Transportation Engineers also produces design criteria for urban roads and residential streets.

Another key reference is FHWA's *Manual on Uniform Traffic Control Devices,* which defines the installation and maintenance of signs, markings, and signals associated with our roadways. The manual describes requirements and recommendations for the

FIGURE 3-4 The view from the Rowena Crest shows the hairpin curve on the Historic Colombia River Highway in Troutdale, Oregon. Photo by Dennis Adams, courtesy of National Scenic Byways Online (www.byways.org).

application and design of traffic control devices, navigational and warning signing, pavement markings, and work zone traffic control devices. Adherence to the manual is a requirement by law.

Since passage of the Intermodal Surface Transportation Efficiency Act in 1991, states have been allowed to develop highway design standards outside of the Green Book criteria. As a result, each state has published its own manual for road design.

Impact of Design Speeds and Level of Service on Roadway Design

Design speeds and level of service are two of the major factors that determine what kind of road and how many travel lanes will be developed for a given project, green or otherwise.

Design Speed

The design speed is the highest speed at which a motorist can drive a road safely under ideal pavement and weather conditions. Consistency and driver expectations are also important factors for selecting an appropriate design speed. Engineers usually set design speeds higher than the anticipated speed limit to allow for the possibility that the speed limit will be higher than anticipated, to provide a margin of safety for motorists exceeding the speed limit, or both. Design speeds are directly related to roadway classification, so once the functional classification of a particular roadway is established, so too is the range of design speed.

Design speed is the single most influential choice designers make. The choice of a design speed has a significant impact on the physical design of a roadway. For example, a higher design speed requires wider shoulders, bigger curves, and larger clear zones. Lower design speeds allow for greater flexibility in the road design.

Transportation engineers have traditionally adhered to the principle that faster is better, so the tendency is to set the highest design speed possible. Creating a green roadway requires a different philosophy for design speed. Instead of maximizing speed, emphasis should be on a design speed that allows the best road for a given location. For example, lowering the design speed may allow the creation of a parkway or boulevard instead of an expressway. It also allows for the implementation of other elements that will improve aesthetics and reduce environmental impacts.

Level of Service

Level of service (LOS) is one measure of user satisfaction with an intersection. LOS is a qualitative assessment that reflects the quality of flow as measured by a scale of driver satisfaction, and is determined by standards set forth in the FHWA's *Highway Capacity Manual*. No matter how innovative, aesthetically pleasing, or environmentally friendly a design is, it has no chance of being constructed unless it meets an acceptable LOS.

The AASHTO Green Book does have some recommended thresholds for LOS, but this measure is usually determined by state policies and influenced by the functional classification of a road. The *Highway Capacity Manual* provides definitions of LOS based on spatial and delay measurements.

An acceptable LOS is determined for each individual transportation project. One general rule of thumb, though, is that each road should have a "passing grade," meaning a C or better, which indicates

FIGURE 3-5 The Swannanoa River Bridge is a concrete box girder bridge that was built in 1966. Located in North Carolina, it crosses the Swannanoa River and Interstate 40. Image courtesy of the Library of Congress-HABS/HAER.

low levels of traffic congestion, minimal accident risk, and the accomplishment of other transportation goals.

Importance of Safety

Safety is a top priority for every transportation project. Although we want to improve aesthetics and develop roads that are more environmentally sustainable, compromising on safety is not an option. Green roadways can be just as safe, if not safer, than traditional roads.

Highway fatalities have dropped significantly over the past decade. Safety has been improved through better vehicle technology, smarter road designs, and reformed behaviors, such as reduced drunken driving. Many accidents are a result of user error, or inappropriate behavior. For example, in 2008, nearly six thousand people were killed as a result of distracted drivers talking on cell phones or texting.

In the United States, well over half of traffic deaths occur on rural roads, though only a quarter of the population is rural. That has a lot to do with narrow shoulders, steeper adjacent slopes, and higher design speeds.

An important part of creating a safe road is to provide a clear zone. The clear zone is the unobstructed area beyond the edge of the traveled way that allows for the recovery of errant vehicles. The width of the clear zone is influenced by several factors, the most important of which are traffic volume, design speed of the highway, and slope of the embankments. The AASHTO *Roadside Design Guide* is a primary reference for determining clear zone widths.

Design Exceptions

One common misconception is that it is not possible to build a road that looks good and fits the neighboring landscape and also is safe and meets accessibility requirements. As a general rule, green solutions can be produced within existing rules and regulations. For example, Paris Pike, which is considered to be Kentucky's most difficult and landmark context–sensitive roadway project, was completed without any design exceptions.

Design exceptions are recommended only in unique situations, such as protecting the visual character of an historic road by allowing trees within the clear zone, or changing the width of a standard median to allow for stormwater management. When a design exception is requested, documentation is typically prepared by the design engineer and forwarded to the state's Design Exceptions Committee and then forwarded to FHWA for approval. FHWA requires that all exceptions from accepted guidelines and policies be justified and documented in some manner. Most states shy away from design exceptions because of the concern that they increase liability.

In Maryland, where Thinking Beyond the Pavement/Context-Sensitive Design originated, staff note that the number of design exceptions has not increased even with more flexible transportation design.

Right-of-Way

A right-of-way is the narrow strip of land that houses a road. As a general rule, the idea is to acquire just enough right-of-way to accommodate a specific transportation project. Wider rights-of-way provide more flexibility for locating a road and more opportunity for landscaping, grading, and structural elements along the road, but they also have more impact on surrounding land uses. Green roadways don't necessarily need wide rights-of-way to be successful, though a creative, multiuse approach to the right-of-way is often the best solution.

Right-of-way costs often represent the single largest expenditure for a transportation project. Right-of-way acquisition for transportation projects typically occurs in one of two ways: (1) right-of-way is acquired once the project engineering is complete; or (2) right-of-way is acquired first, and the project is engineered at a later date to fit within the purchased corridor. This second approach is referred to as land banking, and it can be used for the entire needed right-of-way of a project or for purchase of strategic sections only.

Right-of-way can be acquired either from willing sellers or through eminent domain. With a willing seller, the state negotiates with a landowner to come up with a fair price. This is certainly the preferred approach because everyone is usually pleased with the results. Eminent domain involves the state "taking" the land for the public good. With eminent domain a landowner has no option but to sell his/her land at what is deemed to be a fair price. If eminent domain has to be used, the time it takes to complete a transportation project increases significantly.

It is important to find the balance between the rights of property owners and the responsibility of providing adequate infrastructure. Wider rights-of-way can quickly get expensive, especially in areas where the land values are high. The key is to acquire sufficient right-of-way to accommodate all elements of a transportation project, including utility accommodations, clear zones, drainage ditches, sidewalks, buffer strips, curbs or berms, shoulders and bicycle lanes, motor vehicle travel lanes, and medians.

In urban areas, rights-of-way always seem to be too narrow to accommodate all of the design elements needed to improve connectivity, enhance walkability and community character, and protect existing natural and cultural resources. One approach is to use rights-of-way for a combination of needs, such as trails, roads, utilities, and drainage. One recommended option is to look at multiple uses for the right-of-way. It is possible to use these rights-of-way for stormwater management, as linear parks complete with hiking and biking trails, for active recreation such as ball fields and playgrounds, or even for harvesting biofuels such as switchgrass. The basic idea is to stop thinking of a right-of-way as no man's land that is to be used only to house a roadway.

In the United States there are more than sixty million acres of land within highway and road rights-of-way. That is a lot of land not being used efficiently. If we think outside the box, those sixty million acres can be used for other activities as well.

Typically, a roadway does not use all of the land within a right-of-way, especially near interchanges. This land can be used for a variety of other uses, including passive and active recreation, stormwater management, wildlife habitat and movement corridors, trails, public art displays, bioharvesting, and other uses. Some cities allow parking along roadway rights-of-way during festivals and special events.

In South Dakota, public road rights-of-way are open for the hunting of small game and waterfowl. Texas experimented with growing hay within rights-of-way, but has stopped the process because thieves were stealing the baled hay. The City of Sammamish, Washington, allows the temporary private use of the public right-of-way for events such as block parties, parades, and dumpster placement. A permit fee of $450 is required for each use.

For the Houston Low Impact Development Green Roadway Design Challenge, the AECOM design team used different varieties of switchgrass that vary in color, texture, and height along the edges of the right-of-way to add visual interest. The switchgrass can later be harvested and used as biofuel.

In urban areas, transportation and utility rights-of-way are often used for trails. Introduction of public trails along these corridors requires cooperation and planning between the trail or city agency and a department of transportation (DOT) entity to determine operation and maintenance impacts. Traditionally, DOTs took the approach that transportation rights-of-way were off-limits to any other activity. One reason was concern about the liability issues the DOT would face if someone got hurt in the right-of-way. A number of projects around the country have allowed the construction of multiuse trails within the right-of-way of controlled-access highways. These include:

- I-205 in Portland, OR
- I-70 in Glenwood Canyon, CO
- I-275 in Detroit, MI
- NY Route 104 in Webster, NY

- I-384 from East Hartford to Manchester, CT
- Palisades Interstate Parkway, NY
- Bronx River Parkway, NY
- Hutchinson River Parkway, NY
- NY Route 9, Westchester County, NY
- I-66 in Arlington, VA
- I-90 bridge across Lake Washington in Seattle, WA

For most of these projects, the trail is located in the highway right-of-way but is separated by access control fencing.

Greening Roadway Components

Each transportation project is unique, but projects typically include the same basic design elements. The key is how these elements are combined. For green roadways, these elements are integrated in a way that is respectful of a place and the people who live there.

Travel Lanes

Travel lanes are the actual lanes of the road on which we drive. The number of travel lanes varies depending upon available space and anticipated number of vehicles. The number of lanes needed for a facility is determined during the concept stage of project development. Historically, the highway design process has been based upon the idea that the more travel lanes the better. In fact, even when we build two-lane or four-lane roads, a corridor is often cleared to accommodate future lanes as well. The impact of this larger footprint is significant.

The width of travel lanes is determined by the physical size of cars and trucks. Twelve-foot travel lanes are commonplace for roads with high design speeds. Nine- to ten-foot-wide lanes are appropriate on low volume roads in rural and residential areas.

For green roadways, as a general rule, less paving is better. That means reducing the width of travel lanes, using pervious alternatives for shoulders and parking areas, and emphasizing alternative modes of transportation rather than just building more roads because that is what we have always done.

Effective pavement design is a critical part of every green roadway project. Paving is often selected for either its functionality or its aesthetic quality, but with new materials and construction processes, it is possible to achieve both. For travel lanes, emphasis is very clearly on functionality. For some walks, pedestrian crossings, medians, and other hardscape elements that are highly visible, aesthetics are more of a concern.

The condition and adequacy of travel lanes is often judged by the smoothness or roughness of the pavement. Deficient pavement conditions can result in increased user costs, travel delays, braking, fuel consumption, vehicle maintenance repairs, and crashes.

Most travel lanes are constructed at ground level, but in certain locations they may be either elevated or depressed in an effort to minimize impacts.

Elevated Roadways

Elevated roadways have become commonplace in dense, urban areas, especially where the topography is relatively flat, because they help eliminate conflicts with other roads and can help provide a greater LOS. Elevated roads are expensive and take a long time to build, and they have a major visual and physical impact on adjacent communities.

Most of the elevated highways in this country seem to be mainly a monument to engineering prowess. Elevated highways often are opposed by community groups because they

- create a physical barrier and divide neighborhoods
- promote thru traffic
- are often unsightly
- increase and magnify noise generated by traffic

Efforts to minimize the negative impacts of elevated highways are akin to the old expression about

FIGURE 3-6 The Old Frankfort Pike is considered one of Kentucky's most scenic byways. It passes through six different districts and by four National Historic Register properties on the 16.9-mile drive that runs from Lexington to Frankfort. Photo by A. Crane, courtesy of National Scenic Byways Online (www.byways.org).

FIGURE 3-7 Pedestrians in Venice, Florida, are using a curb extension that improves the visibility of pedestrians and reduces their exposure to motor vehicles. Image courtesy of www.pedbikeimages.org / Dan Burden.

putting lipstick on a pig. Colors and materials can be selected that are visually more consistent with surrounding structures, and landscape massing can be used to help hide part of the highway. Some communities have decided the best approach is to remove elevated highways. When Manhattan's West Side Highway was closed in 1973, the elevated freeway was replaced with new medians, a waterfront park, and a bicycle path to a neighboring surface street.

From a green standpoint, elevated highways, such as those used in Glenwood Canyon and Snowmass Canyon, Colorado (see the case study on both in chapter 8), can be an effective way to build a road in an environmentally sensitive area while reducing the construction footprint.

Depressed Roadways

Depressed roadways are an option when there are major conflicts between land uses, and changes in grades make this a viable option. The basic idea is to maintain activities at the ground level and lower the road. Seattle's new Alaskan Viaduct and Boston's Big Dig (see case study in chapter 6) are examples where the roadway is built belowground. Two major advantages of depressed roadways are that they minimize conflicts with pedestrians and free up land for other uses, such as public open space. The downside is that depressed roadways are very expensive and take longer to construct.

Shoulders

Shoulders are located along the edge of the roadway and provide a place for vehicles to pull off the road if need be. In rural areas, highway shoulders often provide a separate travel way for pedestrians, bicyclists, and other similar users.

The width of shoulders is usually determined through an assessment of pedestrian, bicycle, and motor vehicle needs, and available right-of-way. Shoulder widths vary considerably, from as little as two feet on minor rural roads to between ten and twelve feet on major urban roads. In rural areas, a four-foot-wide shoulder is considered the minimum, with six- to ten-foot-wide shoulders being preferred from a traffic safety standpoint. For many roads, eight feet seems to be a popular width.

In areas where pedestrian volumes are low, or where both traffic volumes and speed are low, a paved usable shoulder can provide pedestrian accommodation.

One concern is that wide, paved shoulders increase the physical and visual footprint of a road. This is a potential problem, especially in areas where visual resources are a priority, and in historic areas where travel lanes were traditionally fairly narrow and the road had no shoulders.

Shoulders may be surfaced for either their full or partial widths. Some of the commonly used materials include crushed gravel, shell, crushed rock, mineral or chemical additives, bituminous surface treatments, and various forms of asphaltic or concrete pavements.

Traditionally, the shoulder would be constructed of the same material as the road because that was the most cost-effective approach. Grass shoulders that are structurally reinforced with geotextile fabrics or other soil supports can be used to limit the perceived width of the roadway and still provide a breakdown area for motorists. One approach that has gained popularity in recent years is to use four-foot paved shoulders for bicyclists, plus six feet of stabilized grass shoulders beyond to minimize the visual impact of a road.

For the Paris Pike project in Lexington, Kentucky (chapter 8), the idea was to reduce the visual footprint of the road. To do this and still provide adequate space for motorists to pull to the side of the road, the shoulders are grass on top of reinforced geotextile fabric.

Curbs and Gutters

Although shoulders are recommended in rural areas, they often are replaced along urban and suburban streets by curbs that help protect an adjacent sidewalk. Curbs help control the movement of vehicles, control access, and help direct stormwater runoff.

There are basically two types of curbs: barrier and mountable. Barrier curbs are typically six to eight inches tall and are used in urban areas on roads with a design speed less than forty miles per hour. Mountable curbs have an angled edge that can be driven over; they are frequently used to allow service vehicles or delivery trucks to access areas while restricting normal traffic. It is common to see curbs, gutters, and drain inlets used together as a way to control stormwater runoff.

Curbs can be constructed from a variety of materials, including concrete, asphalt, and cut stone. Most curbs are constructed of concrete because the material is very durable, will not deteriorate if run over by a vehicle, and can be easily formed to follow the shape of the horizontal roadway alignment. Asphalt, on the other hand, does not hold up well as a curb, and can flatten if run over. In Atlanta, most curbs for city streets are constructed of granite because the material is commonly found within the region. The result is a cleaner, more regal roadway edge that is different from what you find in most parts of the country.

When curbs are used, they typically occur on both sides of the travel lanes, and the road itself is crowned in the middle so all water flows to the edges of the road. Gutters are typically constructed along with the curbs to direct the stormwater runoff to drainage inlet grates or to openings in the curb that lead to bioswales within the right-of-way. Green roadways focus on more natural stormwater management processes that are intended to emulate natural systems that slow down the water and improve water quality. (See "Water Resources" in chapter 9.)

FIGURE 3-8 Rain gardens, like this one in Portland, Oregon, are used for stormwater management. They allow rainwater to soak into the ground instead of running into storm drains or surface water. Image courtesy of the City of Portland.

Medians

The primary function of medians is to improve safety by separating traffic streams and guiding turning movements at intersections. They are often used to accommodate left-turn lanes, which helps improve LOS and reduce potential accidents. Medians can also improve safety for pedestrians by giving them a safe refuge when trying to cross wide roadways with multiple travel lanes. When it comes to creating green roadways, one concept is to consider medians as the "land between the roads," and to consider possible uses of that land.

Medians break up the visual impact of a roadway, especially those with multiple travel lanes. As a general rule, the wider the median the greater the perceived separation between the lanes. If a median is planned, it should be wide enough to have a visual impact upon the road.

Medians can also be landscaped to improve the visual character of a roadway. For residential areas where slower speeds are posted, wider medians can be planted with street trees that create a shade canopy, helping to define the visual character of the neighborhood.

Some medians are depressed and are used as bioswales and rain gardens to help manage stormwater runoff. The travel lanes are graded so they drain toward the middle, into the median. The stormwater can be managed and treated, and delaying the runoff will reduce the peak flow of a given storm, helping reduce the likelihood of a wastewater system overflow. This is an effective approach for green roadways.

The most memorable parkways in the United States are those that have large medians with trees, wildflowers, native grasses, and rolling hills to create an area that has aesthetic appeal in and of itself. Medians should be viewed as opportunities to enhance the driver experience in addition to creating a safer roadway. They can be used to enhance aesthetics, manage stormwater runoff, provide wildlife habitat, and even grow switchgrass biofuel for energy generation.

Another benefit of using a wider median to separate travel lanes is that each set of lanes can have independent horizontal and vertical curves, and this approach allows the road to better fit existing landforms. Wider medians can also be used to protect existing cultural or natural resources; the road basically splits to avoid the resources.

The downside of using medians is that they expand the footprint of a given road, potentially resulting in a greater impact on adjacent resources. In general, wide medians should be used where there is sufficient space to accommodate the larger roadway footprint while still allowing for other uses along the edges of the right-of-way. Medians improve safety, make the roadway more visually appealing, and can be used for stormwater management.

Intersections

An intersection is an area where two or more streets join or cross at the same level. The efficiency, safety, speed, cost of operation, and capacity of the highway system depend on the design of its intersections. There are three types of intersection control: stop, signal, and roundabout. According to FHWA, of the more than three million intersections in the United States, approximately three hundred thousand of them are signalized. Most signalized intersections are in urban areas; only about 10 percent of intersections in rural areas are signalized.

Right-turn lanes and left-turn lanes are typically included at heavily used intersections. Uncontrolled intersections are those that lack signs or signals. In the United States uncontrolled intersections are greatly frowned upon because of safety and liability concerns.

Types of Interchanges

Intersections are often visual landmarks that function as community gateways. As a result, the aesthetics of intersections are often a priority. Using different textures

FIGURE 3-9 Along the Easter Parkway in Louisville, Kentucky, a trail runs through the middle of a median between travel lanes. Image courtesy of Louisville Metro Parks.

and colors for paving help highlight crosswalks and islands. Native plantings integrated with bioswales can help improve visual quality and create a healthier environment. Adding way-finding signage and public art helps make intersections more informative and more exciting. In Houston's Galleria District, large stainless steel ovals hang over each intersection. The ovals include the streets names, and collectively they create unique visual landmarks for the Galleria area.

There are a number of different approaches to designing interchanges:

At-grade intersections are areas where two or more roadways join or cross at the same level. They are one of the most critical and most complicated elements in highway design and present serious safety challenges because of the conflicts that occur among motorized vehicles, and between motorized vehicles and pedestrians/cyclists. At-grade intersections work best at lower speeds. At higher speeds, such as along interstates, every effort is made to eliminate at-grade intersections.

Diverging diamond interchanges improve traffic flow by eliminating left turns. The catch is that they do so by having drivers crisscross at each end of the interchange as traffic lights let one side go, then the other. The process may seem strange, but traffic engineers say that these types of interchanges work.

Median U-turn crossovers are at-grade intersection designs that replace each left turn with a U-turn.

They are frequently called Michigan lefts because they have been used along Michigan roads and highways since the late 1960s.

Double teardrop interchange is a unique double roundabout configuration. It increases capacity with fewer approach lanes. The double teardrop interchange improves traffic flow, slows down traffic, and reduces accidents.

Channelized intersections use pavement markings or raised islands to designate the intended vehicle paths. The most frequent use is for right turns, particularly when accompanied by an auxiliary right-turn lane.

A **roundabout** is a channelized intersection with one-way traffic flow circulating around a central island.

Flared intersections expand the cross section of the street to provide space for left-turn lanes.

Horizontal and Vertical Curves/ Alignments

Horizontal and vertical alignments establish the general character of a roadway. A horizontal curve in a roadway refers to the alignment, or how straight the roadway section is. A vertical curve refers to a roadway's change in elevation, or the flatness of the

FIGURE 3-10 The Ottawa River Parkway is an interchange that passes along the Ottawa River in Ottawa, Ontario, Canada. Besides being a scenic route, it also serves commuters who work in the downtown area. Image courtesy of Wikipedia Commons.

roadway. Alignment constraints typically include topographical variation, natural resource areas, property ownership, land use, cost, and environment.

The AASHTO Green Book includes a set of design curves for both horizontal and vertical alignments.

Horizontal Alignment

The configuration of line and grade affects safe operating speeds, sight distances, and opportunities for passing and highway capacity. The horizontal alignment for a specific road has a significant impact on potential accidents, how well the road fits existing conditions, and the overall visual quality of the final project.

Horizontal alignments consist of a combination of tangents, which are straight sections, and curves. A number of different types of curves make up horizontal curves, including the following:

Simple curve–A simple curve, often referred to as a circular curve, has a constant circular radius, and is the most frequently used curve because of its simplicity for design, layout, and construction. Generally, simple curves should be used on most projects. The radius for horizontal curves is measured to the horizontal control line, which is typically the centerline of the alignment.

Broken back curves–Broken back curves have short tangents between two curves turning in the same direction. They are awkward and unsafe, and should be avoided. There really is no excuse for a new transportation project to use broken back curves. They are also called compound curves.

Spiral curves–Spiral curves are generally used to provide a gradual change in curvature from a straight section of road to a curved section. Spiral curves are considered better design than compound simple

FIGURE 3-11 The Old Canada Road Scenic Byway passes Wyman Lake near the town of Bingham, Maine. The byway follows Route 201 for seventy-eight miles from the town of Solon to the Canadian border. Photograph courtesy of Maine Office of Tourism.

curves. One benefit of a spiral curve is that it improves the appearance of circular curves by reducing the break in alignment perceived by drivers. The result is a road on which smooth movements create a sense of rhythm for the driver. Some states allow the use of spiral curves, but others, such as South Dakota, Wisconsin, and California, do not. The Green Book does not preclude the use of spiral curves, but some states prefer not to use them because they are more difficult to stake out during the construction process. On the other hand, states such as Ohio, Montana, and Kentucky use spiral curves frequently. Oregon actually requires spirals to be used on curves greater than or equal to 1 degree. Kentucky's Paris Pike was designed with spiral curves, and the curvilinear design seems to fit the rolling hills and lush horse farms. You get the feeling you are driving with the landscape, not through it.

Vertical Alignment

The vertical alignment of a road refers to the rise and fall of a road, or how it will follow the lay of the land. Vertical alignments are controlled by design speed, topography, traffic volume and composition, highway functional classification, safety, sight distance, typical sections, horizontal alignment, climate, vertical clearances, drainage, economics, and aesthetics.

A number of factors influence the vertical alignment of a highway, including

- natural terrain
- minimum stopping sight distance for the selected design speed
- the number of trucks and other heavy vehicles in the traffic stream
- the basic roadway cross section, i.e., two lanes versus multiple lanes
- natural environmental factors, such as wetlands, and historic, cultural, and community resources (AASHTO, 2004)

Vertical alignments should fit the existing landscape as much as possible, and be designed to minimize cut and fill. Frequently, emphasis is on balancing cut and fill so no additional soil has to be brought to, or carried from, the site. There is an old saying that "moving dirt is cheap and easy, but hauling it is not." When horizontal and vertical alignments are

Figure 3-12 The gently curving road appears to be carved into the contours of the land, providing a gradual change to the driver, who is met with constantly changing views. Image acquired from depositphotos.com.

designed separately, the result can be extensive cuts and fills that are unsightly and negatively impact existing natural resources.

Superelevation is the banking of a roadway around a curve. The basic idea is to counterbalance the outward pull of a vehicle traversing a horizontal curve. This approach allows a vehicle to safely travel through a curve at a higher speed than would be possible with a normal road section. Superelevations are used primarily on high speed roads.

Coordinating Horizontal and Vertical Alignments

The process of incorporating horizontal and vertical elements into a roadway's design begins with the identification of the proposed corridor and the location of critical constraints that must be considered for preservation throughout the design process. The best horizontal and vertical alignments are ones that follow the natural contours of the land. Both alignments should be designed to minimize impacts and enhance attractive scenic views. This is important for green roadways.

A road is typically shown in cross section because this is often the best way to see how it will fit with the surrounding landscape. The cross section of a road usually shows the travel lanes; roadway, including shoulders; median; bicycle and pedestrian facilities; utility and landscape areas; drainage channels and side slopes; and clear zone.

Coordinating horizontal and vertical curves is one of the best ways to develop a road that fits the

FIGURE 3-13 The Upper Delaware Scenic Byway in New York State is a seventy-mile stretch along the Delaware River. Image courtesy of New York Department of Transportation.

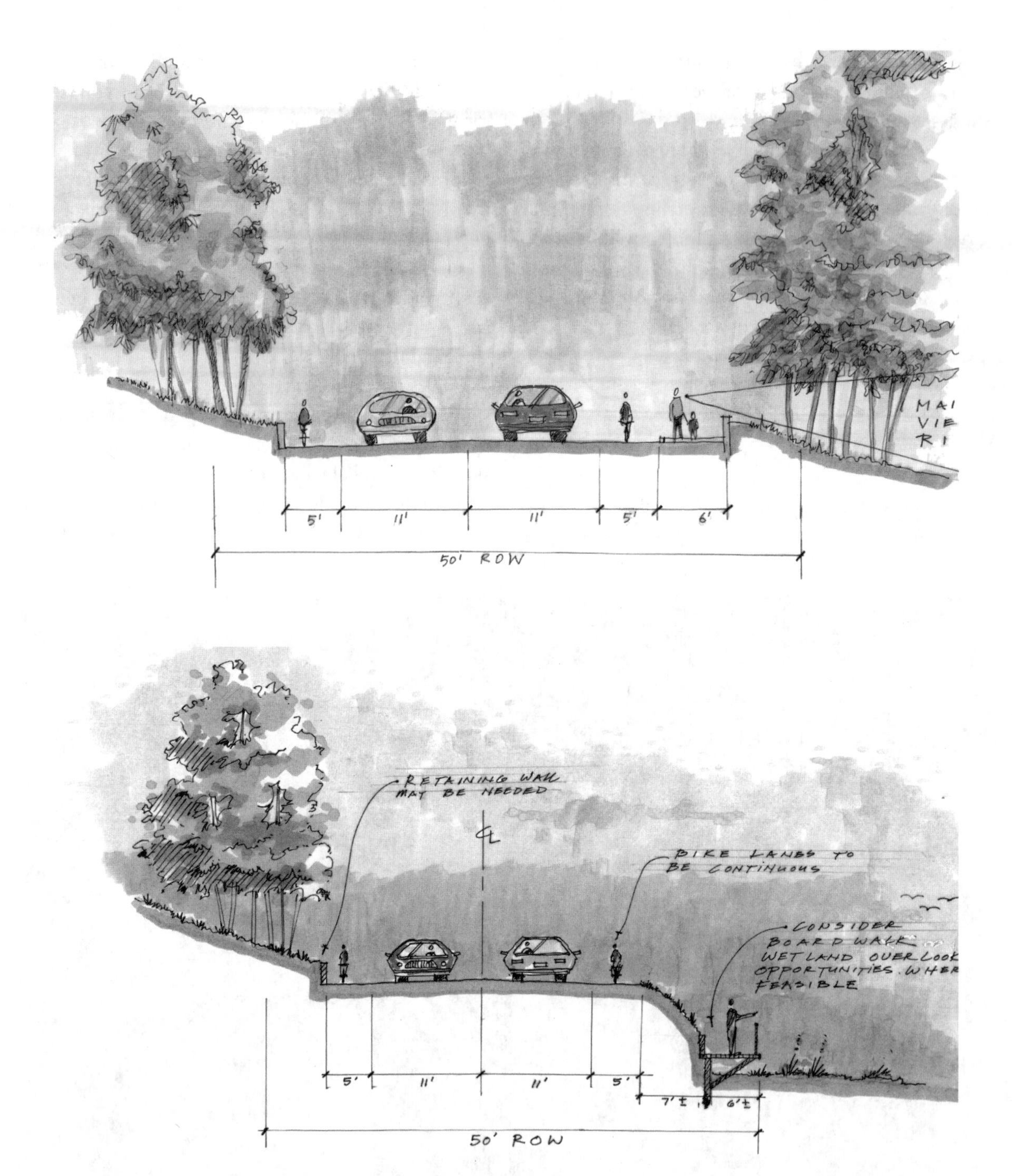

FIGURE 3-14 The right-of-way for five alternative potential roadway cross sections for Riverside Parkway in Roswell, Georgia, varies in width from 50 to 120 feet. Wider rights-of-way provide a greater opportunity to include bike lanes, separate multiuse trails, wider shoulders, and landscaped medians. Images courtesy of EDAW.

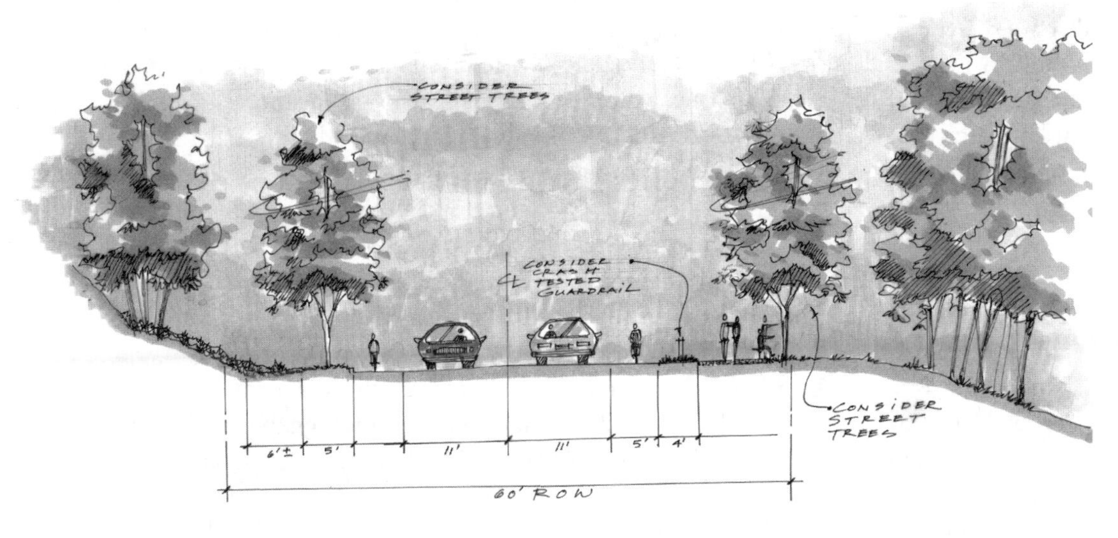

CONSIDER STREET TREES
CONSIDER CRASH TESTED GUARDRAIL
6'±
5'
11'
11'
5'
4'
60' ROW
CONSIDER STREET TREES

LANDSCAPE MEDIAN
6'±
6'±
5'
11
10'±
11'
5'
6'±
10'±
CONSIDER CURB
CONSIDER BENCHES AT RIVER VIEWS
80' ROW

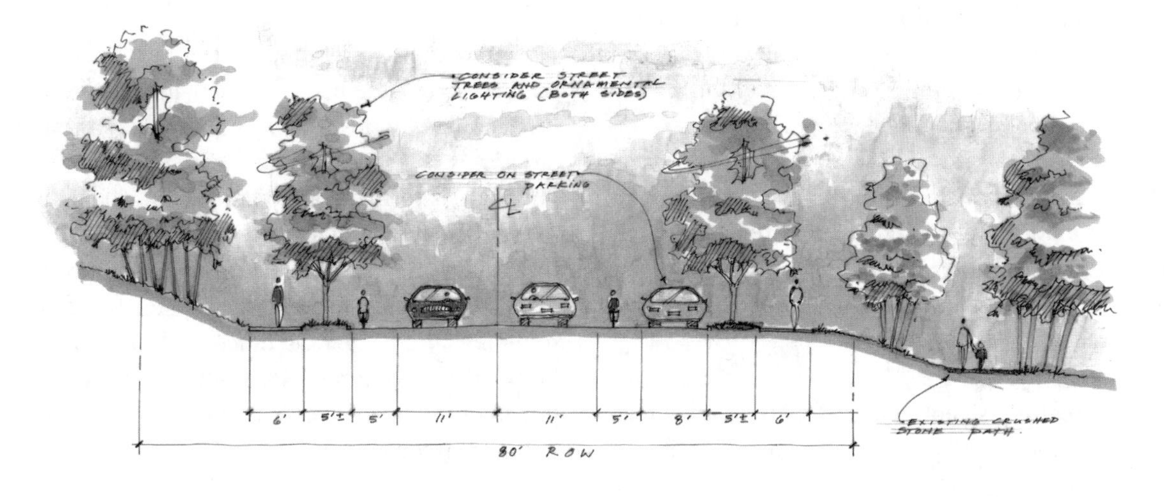

CONSIDER STREET TREES AND ORNAMENTAL LIGHTING (BOTH SIDES)
CONSIDER ON STREET PARKING
6'
5'±
5'
11'
11'
5'
8'
5'±
6'
80' ROW
EXISTING CRUSHED STONE PATH

landscape. Early decisions about the alignments of both have a significant impact on the amount of cut and fill involved, and the resulting overall quality of the proposed roadway, and how it fits with the adjacent landform. As a general rule, the amount of grading should be minimal to protect existing natural resources.

Alignments that follow existing contours generally fit the existing landscape pretty well. In contrast, alignments that are perpendicular to contours result in the greatest changes, including major cuts and fills. Changes to existing landforms should be done in concert with horizontal and vertical alignment of the roadway, and should be consistent with existing natural patterns.

It is important that road projects in mountainous areas include a multidisciplinary team of civil engineers, geotechnical engineers, and landscape architects to ensure that the inherent character of a hillslope's natural bedding planes, fractures, joints, and overall stability are considered, and that rock cuts are designed to avoid the need for rock fall protection fencing whenever possible (Design Workshop, 2006).

Some general recommendations for grading that is consistent with a green roadway approach are as follows:

Create artful earthwork that fits the surrounding landscape. Grading is the foundation of all aspects of the corridor. In addition to grading for effective roadway alignment, carefully consider contour grading. Create landforms that respond to the uniqueness of the site, the surrounding landscape, and the roadway travel experience.

Blend earthwork with existing slope conditions. Patterns of topography should be considered with proposed grading. Valleys, high points, and ridges require graded transitions, rather than abrupt embankment cuts or fills.

Avoid creating steep slopes. Smooth, moderately inclined slopes will blend more readily with the surrounding landscape, are safer to maintain, and are less vulnerable to erosion. In addition, flatter slopes reduce the need for guardrails and provide better accident recovery in the vehicle clear zone. Where site conditions and cost analysis permit, acquire adequate right-of-way to provide enough land to design and build the desired slope and grade. In some locations, steeper slopes may be unavoidable to protect important natural or cultural resources adjacent to the highway.

Create smooth landform transitions and revegetate slopes. Finish-grading techniques such as slope rounding at the top and bottom of cuts should be used to create smooth landform transitions that blend with the natural terrain. Carefully grade slopes around natural outcrops and abrupt topography to improve aesthetics and allow for easier and more cost-effective maintenance. Ensure that all constructed slopes are revegetated. In addition, soil-coloring treatments that blend newly cut or filled soil with existing soils should be implemented.

Do grading via plan views as well as sections. The combination of the two provides a more sensitive grading solution that fits the existing terrain. Using standard computer-aided design cut-and-fill applications is an acceptable way to develop an initial grading plan, but the final grading plan needs to adjusted by hand to create an aesthetically pleasing landscape.

Design rock cuts to be natural in form. Ensure that rock cuts are designed to look natural in form, texture, and color in relationship to the surrounding landforms. Customize fracture rock cuts to match natural rock form and use naturalized bedding planes to avoid creating a sheer, unnatural rock face. Ensure that all designed landforms are natural in appearance and blend with the topography and geology of the surrounding landscape. Match new rock and soil excavations with existing rock and soil using rock staining, soil-coloring treatments, and/or accelerated weathering techniques. Such treatments will successfully blend newly cut or filled soil and rock with existing weathered rock. Where site conditions and cost analysis permit, acquire adequate right-of-way to provide enough land to design and build the desired rock cut slope and grade (Design Workshop, 2006).

Traffic Calming

Traffic calming devices are simple street design features that cause motorists to drive with more care, to drive more slowly, or perhaps via another route. Traffic calming is most often applied to existing streets where the objective is to reduce vehicle speed. Traffic calming appears to be one of the more cost-effective ways to promote pedestrian and bicycle use in urban and suburban areas, where walking and bicycling are often hazardous and uncomfortable. Traffic calming measures also increase driver attentiveness, so accidents are less likely, making the roads safer for all users. They also help slow down traffic enough to allow for concepts such as complete streets (www.completestreets.org), which focus on streets that are safe and accessible for pedestrians, transit riders, bicyclists, and drivers.

The majority of traffic calming devices make slight alterations to the street's geometry, reducing its real or perceived width, or causing the driver to negotiate curvature or pavement texture.

Traffic calming measures are used most often in urban areas, suburban town centers and villages, suburban high density areas, and rural villages. They are not appropriate for freeways and expressways, and are seldom used in rural areas.

In Delaware, 67 percent of the households and businesses on a road must sign a petition for traffic calming devices to be installed. The Delaware DOT makes an initial determination of eligibility for traffic calming and then gathers acceptable field data. The department then rates eligible traffic calming projects, and the highest ranked projects are programmed for installation, subject to available funding (Delaware DOT, 2000).

Some of the most common traffic calming devices include medians, traffic islands, speed bumps, traffic circles, roundabouts, narrow lanes, double teardrop interchanges, chicanes and horizontal lane shifts, pedestrian-scale elements, textured paving and rumble strips, and diverging diamond interchanges.

Medians–As discussed previously, medians provide a physical separation between opposing traffic flows. They help improve safety, provide a refuge area for pedestrians seeking to cross the road, provide space for vehicles making left turns, and can also improve aesthetics. For some projects, it is more important to minimize the footprint of the right-of-way and road construction to protect important cultural, historical, or natural resources along the corridor. In these situations, travel lanes and road shoulders should be kept to a minimum, and medians should not be used.

Traffic islands–Traffic islands help direct the movement of vehicles at intersections. These can either be raised via the use of curbs or be painted directly on the roadway. Traffic islands are commonplace, and they are one of the most successful ways to simplify movements at intersections.

Speed bumps–Speed bumps are typically made of asphalt, concrete, rubber, concrete pavers, or other material that runs perpendicular to a road and is intended to slow down traffic. Although some transportation departments promote speed bumps as being "safe and comfortable for automobiles," most drivers would tell you they are jarring and disruptive (Delaware DOT, 2000). A typical speed bump is around three inches high, but on private roadways you can find some as high as six inches. If designed correctly, streets should not need speed bumps to slow traffic. Raised crosswalks serve the same basic function as speed bumps, but they are typically higher and wider, and extend from sidewalk to sidewalk. They can be effective for residential areas where slow speeds are desired.

Traffic circles–Most traffic circles have circular center islands and circular perimeters formed by intersection corners. The typical circle has a center island with two levels: a base that is mountable by an automobile and a raised center.

Traffic circles are usually most effective when deployed at three-way or four-way intersections, and when they are constructed in a series. They are sometimes located in the middle of the block. Circles reduce motor vehicle speeds and result in a big reduction in the number of accidents. Circles reduce crashes by 50–90 percent when compared to two-way and four-way stop signs and traffic signals by diminishing the number of conflict points.

Roundabouts–These are designed to slow entering traffic and allow all the traffic to flow through the junction freely and safely. Roundabouts are basically larger versions of traffic circles. They have significant benefits during off-peak periods, when induced travel would be minimal because the roadway network is generally undercapacity. If there are few conflicting vehicles present, entering vehicles can enter with minimal delay. The average delay at a roundabout is estimated to be less than half of that at a typical signalized intersection. Roundabouts are becoming very popular along many green roadways.

Based on research found in the National Cooperative Highway Research Program's publication *Roundabouts in the United States*, roundabouts result in a 48 percent reduction in total crashes when compared to an urban traffic signal.

Narrow lanes–Narrowing the width of roads helps slow down traffic. A series of small streets yields a better bicycle and pedestrian environment than a hierarchy of a few larger streets, and research has shown that narrow roads are safer than wider roads. Many green roadways with lower speed limits use narrow lanes.

Narrowing the street at an intersection, through the use of curb extensions, is a versatile and widely used traffic calming measure.

Vermont has developed design standards for narrow lane and shoulder widths to help maintain the character of small towns and villages throughout the state. Portland, Oregon, recommends twenty-foot-wide streets for new residential areas. The city's fire department has also found that it is more economical to use smaller fire trucks that fit these narrower streets than to build twelve-foot-wide lanes. Eugene, Oregon, has also reduced its design standards for some roads to only twenty feet.

Double teardrop interchanges–A double teardrop interchange is a unique double roundabout

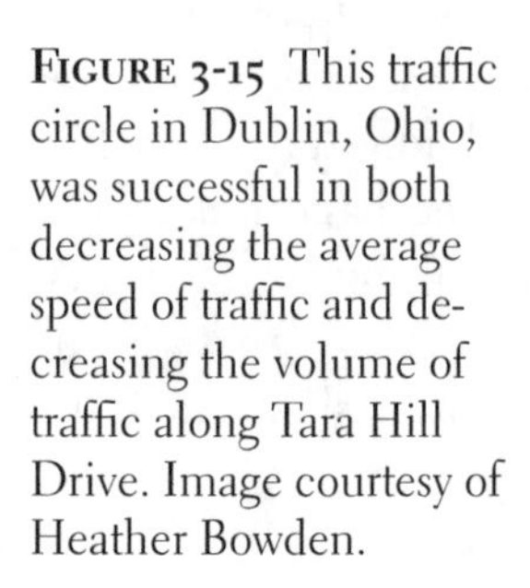

FIGURE 3-15 This traffic circle in Dublin, Ohio, was successful in both decreasing the average speed of traffic and decreasing the volume of traffic along Tara Hill Drive. Image courtesy of Heather Bowden.

FIGURE 3-16 The Old Market Roundabout in Bristol, England, has traffic lights at every entrance (not typical for roundabouts), a road that passes beneath it, and gardens in the center. Image courtesy of Wikipedia Commons.

configuration. It increases capacity with fewer approach lanes. Roundabouts are one of the safest, most efficient intersection control techniques, having fewer, less severe accidents. Motorists are able to navigate the intersection without having to stop at traffic signals, reducing travel times. No power is required for signals.

The double teardrop intersection uses one-third less area than a traditional diamond intersection, preserving existing vegetation and reducing traffic noise. The design eliminates stoplights, which reduces car idling and, as a result, improves air quality.

Chicanes and horizontal lane shifts–A chicane is a trapezoidal island added to the side of a road. Cars have to slow down to maneuver around the chicanes and this helps reduce traffic speed. Horizontal lane shifts involve creating a meandering, curvilinear road that has limited straight tangents. The curves slow down traffic and help create a more aesthetically pleasing roadway.

Pedestrian-scale elements–One of the best traffic calming measures in urban areas is to create a more pedestrian-oriented scale. This includes adding street trees along roads and adding wide sidewalks with street furniture elements such as signs, signals, street lights, walls, fencing, and pedestrian furnishings such as benches, shelters, and trash receptacles. In commercial settings, the placement of buildings directly along the street is a highly effective traffic calming measure. A street lamp height of twelve to fifteen feet supports a traffic-calmed environment by signaling an area of special concern where pedestrians are present. Curbs are important in traffic-calmed settings because they signal a lower design speed to motorists.

Textured paving and rumble strips–Textured paving is used to let motorists know they need to slow down or pay attention, because there is a change in how the roadway functions. Textured paving is often used at crosswalks, interchanges, or other locations where traffic needs to slow. Rumble strips can be used as a traffic-calming measure for local roads. They are typically used to warn about upcoming changes in the road, such as a crosswalk, roundabout, or intersection.

FIGURE 3-17 The combination of narrower streets, on-street parking, and wider pedestrian spaces slows down traffic and creates a more pedestrian-friendly corridor along Atlanta's 15th Street. Image courtesy of EDAW.

Diverging diamond interchanges–This type of interchange is intended to eliminate left turns, which are one of the greatest conflict points for possible accidents on roadways. Traffic going in one direction stops so traffic going the other way can shift over to the two left lanes. Left-turn lanes are eliminated.

Walls and Barriers

Retaining walls, freestanding walls, and barriers are strong visual elements because of their size and proximity to the roadway. Walls are often needed for roadway projects to minimize impacts from excessive grading and minimize views. The key is to make sure that walls are integrated so that they complement the visual character of a community. The color, texture, and pattern of walls have a major influence on a driver's perception of a roadway. For walls, the following recommendations should be considered:

Use grading to minimize wall height. Aesthetic improvements for walls should be considered in concert with specific site characteristics, available space, cost, and noise protection procedures. Where possible, the height of freestanding walls should be minimized to reduce negative visual impacts. Use a combination of earth berms and walls to achieve structural integrity and buffer sound while limiting actual wall height. Walls used only for visual screening should not exceed ten feet. On gradually sloping grades, ensure that the top of the wall transitions appropriately with

FIGURE 3-18 This diverging diamond interchange is part of Utah Department of Transportation's I-15 Corridor Expansion Project. Image courtesy of Utah Department of Transportation.

the slope. Match the top of the wall with the adjacent contour. Use a step or change of plane where retaining walls are getting too tall.

Provide landscape planting and setback space between clear zones and walls. Landscape plantings in front of walls will soften the appearance of large wall faces.

Select a simple design palette. Choose a simple design palette of material, pattern, color, and texture that coordinates with the visual character of a corridor.

Choose an appropriate visual design subject. Use visual design themes and/or pictorial motifs consisting of simple patterns and distinct surface texture, and carefully design the motifs' composition on the wall. Ensure that visual design themes and/or pictorial motifs are an appropriate subject and scale for the highway segment in which they are located.

Create visual breaks and interruptions to avoid monotony along walls. Use staggered and/or curved walls of varying lengths to provide visual interest along extended stretches of sound walls. Battered walls, which are inclined walls, can provide additional interest. Shadow patterns that shift and change throughout the day can be introduced to create visual interest.

Separate walls from other highway structures and set back from travel lanes. When practical, avoid attaching walls to concrete barriers and/or retaining structures. When walls are attached to such structures, avoid mixing materials and incompatible forms.

Use appropriate wall caps. Wall caps that mimic or follow the shape of the background tend to be less obtrusive.

Incorporate design considerations. These include horizontal patterns to deemphasize the height. Vertical textures and patterns will minimize the apparent visual length of the wall. Vary the grade along the face of the wall and integrate with planting to minimize the apparent height of the wall.

Use special finish options on vertical surfaces. Options that can change the visual appearance of

FIGURE 3-19 The Path Most Traveled is an artist-designed wall located along six miles of the Pima Freeway/Loop 101. The artwork reflects the culture and character of Scottsdale, Arizona, as a unique arts community. Image courtesy of Arizona Department of Transportation.

vertical surfaces include form liners, sandblasting, exposed aggregate, pigmented coatings, integral dyes, concrete coatings, architectural veneers, and modular structural units.

Noise and Noise Walls

Noise is an unfortunate part of every transportation project. Noise walls are one way to minimize the problem, but they often create visual barriers that further divide neighborhoods and lessen the driving experience for motorists. Noise needs to be addressed, but the traditional use of noise walls may not be the best approach for green roadways.

Field reconnaissance and map review is undertaken to identify and classify noise receptors that may be affected by a proposed action. A noise study discusses whether noise mitigation appears feasible for the affected properties. The consideration of noise abatement is required for all affected receptors on federally funded and federal-aid projects. Transportation departments typically consider the following noise abatement strategies:

- traffic management measures
- alteration of horizontal and vertical alignments
- construction of noise barriers
- acquisition of property rights for construction of noise barriers
- noise insulation of public use or nonprofit institutional structures

Noise levels are measured in units called decibels. The most difficult sound-blocking problems are associated with low frequencies (thirty to one hundred decibels) such as a loud, pulsating bass.

Noise walls are freestanding barriers built parallel to a highway near residential areas and public facilities. They are installed where traffic noise exceeds or is expected to exceed established threshold levels. Walls typically range in height from ten to twenty feet tall. Noise walls may help reduce noise impacts on neighborhoods, but they also block views and visually separate the neighborhoods they intend to protect. One problem with many walls and abutments is that they try too hard to look natural in very unnatural ways. Sometimes the best approach is to

Figure 3-20 One major problem with sound walls is that they visually block views and create an enclosed corridor. One solution is to use Plexiglas sound walls that still provide the functionality without creating a visual barrier. Image acquired from depositphotos.com.

create a simple architectural element that looks like a wall.

Sound panels–Acoustical sound panels are available that actually absorb road noise instead of just reflecting the noise. Many of these panels are not just "sound-absorbing" but also "sound-blocking," so sound doesn't penetrate all the way through the panel.

Vegetation for noise abatement–As a general rule, vegetation has little if any impact on reducing noise. Dense rows of evergreens may help create a psychological separation, but they do not reduce noise significantly.

Absorptive pavements–Porous or acoustically absorptive pavements can significantly reduce roadside or on-road noise. As a general rule, sound absorption increases with an increase in surface texture. For example, asphalt mixes with smaller aggregates absorb sound better than those with large-size aggregates. Superpave can reduce noise by 6–7 percent, Stone Matrix asphalt by 12 percent, and double layer porous asphalt by as much as 39 percent.

Traffic Barriers

Concrete traffic barriers, metal beam guard fences, and pedestrian control fences are visually prominent features of the highway. The primary purpose of these types of barriers is to prevent a vehicle from leaving the roadway and striking something adjacent to the road.

To achieve the aesthetic goals along a specific highway, the design of barriers, guard beams, and fences should be visually coordinated with other design elements. Color admixtures, chemical staining, painting, acid etching, textures, sandblasting, and faux finishes can all improve the appearance of concrete barriers.

The Texas Department of Transportation, which is known for its innovative transportation solutions, recommends the following design tips that will assist in blending concrete traffic barriers and metal guard beams into the visual context of the highway:

- Barriers can blend or contrast with the background.
- Colors that contrast with the surroundings will be more prominent visually.

Figure 3-21 The use of bold, angular geometric forms painted on jersey barriers helps blur the barrier's edges and soften the visual impact. This approach is intended to induce nearby traffic to slow down, producing a safer, more bike- and pedestrian-friendly thruway. Image courtesy of John Locke.

- Colors that have the same value as the surroundings will blend in.
- Textured forms for concrete traffic barriers can help establish regional themes.
- The smooth, strong lines of these elements can add definition and clarity to the roadway, reducing the visual complexity of the roadway.
- Special themes can be developed in the patterns used on traffic barriers to avoid the look of a universal interstate highway.

Cost is always a concern when selecting a barrier design. Aesthetic barriers might have a higher upfront cost than standard steel barriers and may be more expensive to maintain. Weathering steel guardrails are an example of an inexpensive barrier that may be considered acceptable in certain surroundings. Corten steel is a popular material where a weathered, rustic look is desired. Weathering steel has had durability problems in a few areas.

Road barriers are usually categorized as flexible, semi-rigid, or rigid, depending on their deflection characteristics on impact. Flexible systems are generally more forgiving, but rigid systems are generally more durable.

Flexible barriers–Flexible systems are designed to "give" or break away upon impact. The most common are cable systems that can be used along low-speed roadways. Cable systems use posts driven into the ground at fixed vertical intervals; cables are attached to the posts. They are inexpensive and simple to install and they are relatively inexpensive to repair. One of the biggest advantages of cable systems is that they are visually unobtrusive. Cable systems do have several drawbacks. They can sustain considerable damage in vehicular crashes or by snowplows, and the tension on the cables needs to be adjusted to maintain the proper level.

Semi-rigid barriers–Semi-rigid systems are designed to bend, but not break. Three of the most popular semi-rigid barriers include:

- blocked-out W-beam–uses a heavy post with a block out and corrugated steel face (W-beam)

FIGURE 3-22 The Hehuanshan Road leads to the top of Hehuan Mountain in central Taiwan. Colorful red and yellow guardrails are used to make the narrow, winding road safer, while also adding visual interest. Image courtesy of AECOM.

- blocked-out thrie beam–similar to the blocked-out W-beam guardrail except it has a deeper corrugated metal face. Thrie beam barriers may be aesthetically pleasing to some rural communities because they are less urban in character.
- steel-backed timber rail–is a rustic alternative to the standard metal beam guardrail. A steel plate provides the needed tensile strength, and the wood members provide a rustic appearance.

Rigid barriers–Rigid guardrails that deflect upon impact are generally preferred from a safety standpoint due to the lower impact forces on the vehicle. The most commonly used barriers are made of concrete. There are three types: rigid, rigid anchored, or unrestrained rigid systems. These are commonly used in medians, especially in urban areas.

Jersey barriers are popular in many states, although critics bemoan the negative impact they have upon visual quality and community character.

Stone guardwalls function like rigid concrete barriers but have the appearance of natural stone. These walls can be constructed of stone masonry over a reinforced concrete core wall or of simulated stone concrete.

FIGURE 3-23 The Golden Gate Bridge, a suspension bridge spanning the San Francisco Bay, links San Francisco to Marin County. The color of the bridge, called "International Orange," was selected to complement the natural surroundings and enhance its visibility in fog. Image acquired from depositphotos.com.

Bridges

Each year in this country we spend $6–8 billion for highway bridge design, construction, replacement, and rehabilitation. Bridges are some of the most visible components of a road, helping traverse natural obstacles such as rivers, creeks, and lakes, and human-made obstacles such as other roads and developments. Because of their visibility, bridges can become community landmarks.

Many of the bridges from the golden days of transportation have a much greater aesthetic appeal than many bridges of today. Today's bridges generally fit into one of two categories: (1) landmark bridges that place importance on aesthetics and high design; and (2) utilitarian bridges where the objective was to build a functional bridge for as little money as possible.

Bridges are viewed from two perspectives. Traveling over the bridge deck, the driver of a vehicle sees the travel way, bridge railings, and the view to either side. If the bridge crosses over another roadway, water or land both on its side and underneath can also be viewed from this perspective. Because they are so visually dominant, the aesthetic quality of bridges should be considered when designing a green road. Bridges function as visual landmarks and are an important part of the sense of identity of many major roadways.

FIGURE 3-24 The Millaau Viaduct, the world's tallest bridge, spans the Tarn River and the Tarn Valley in the central-south region of France. At 1,025 feet tall, and 8,071 feet long, the bridge is a striking visual landmark. Image acquired from depositphotos.com.

There is no shortage of examples of bridge projects that successfully address both aesthetics and the minimization of environmental impacts. For the Chanhassen Pedestrian Bridge in Chanhassen, Minnesota, the bridge was designed to be complementary with the adjacent Landscape Arboretum. The landscaping for the bridge was completed in 1997 and has been effective at reducing the visual impact of the structure.

The Vancouver Island (British Columbia) Highway Project consists of more than ninety bridges along 142 miles of highways, including two of the largest that span the Big Qualicum River and the Tsable River. Construction of the bridges was a concern because both rivers provide important habitat for coho, chum, and chinook salmon, as well as steelhead and cutthroat trout. Both bridges were constructed of cast-in-place concrete because the weight of the structure could be greatly reduced, resulting in smaller bridge piers spaced farther apart. To protect fish habitat at the Tsable River Bridge, longer spans were used for the bridge and the piers were set back a minimum of thirty feet away from the riverbank. Another big advantage is that the bridge surface could be built at the top of the piers almost two hundred feet above the valley floor. Because the valley bottom was used only for the delivery of materials, most of the ancient Douglas firs were protected. And during construction of the bridge, work within the riverbank perimeter was restricted to the period between June 15 and September 15 so as not to damage fish stocks (Sipes, 2003).

MATERIALS

A bridge consists of a superstructure and a substructure. The superstructure includes the bridge deck and beams; the substructure includes the cap and foundations of the abutments and the cap, columns, and foundations for bridge piers or support columns.

The design of new bridges requires an understanding of construction materials that offer a greater variety of design options for enhancing aesthetics and minimizing environmental impacts. The Transportation Equity Act for the 21st Century (TEA-21) emphasizes the use of innovative materials that are stronger, more durable, and less toxic to the environment.

Fiberglass bridge structures are being used for pedestrians and equestrians in large part because they are modular in design, lightweight, easy to install, and don't require large footings. Some manufacturers produce bridge kits where no piece weighs more than one person can lift, making it possible to carry a bridge in and install it in an environmentally sensitive area without requiring cranes or heavy equipment.

FIGURE 3-25 The Millennium Bridge is a cable-stayed bridge that spans the Moraca River in Podgorica, Montenegro. Image acquired from depositphotos.com.

High-performance composite materials, such as fiber-reinforced polymers (FRP), are being used for vehicular bridges because the high strength-to-weight ratio makes such bridges easy to manufacture and fabricate. FRP composites are made of fiberglass or carbon fibers, resin, fillers, and additives; fibers provide increased stiffness and tensile strength, resin adds high compressive strength and binds the fibers together, and fillers are used to fill voids and reduce costs. Bridges using composites have been constructed in a number of states. In 1997, the Tom's Creek Bridge was the first bridge constructed by the Virginia Department of Transportation (VDOT) using FRP components. The results have proven so encouraging that the department has since used FRP components on a number of other bridges throughout the state.

Other innovative materials can also help reduce environmental impacts. Paints and primers have traditionally been used on bridge surfaces to prevent rust and corrosion, but many are toxic and some have been found to be carcinogenic. An alternative approach is to use a coating process based on ionic self-assembled monolayers, which can be used on steel, aluminum, and other metals and alloys. These monolayers work at the molecular level and can inhibit corrosion without using hazardous materials.

With new bridge-building materials, industrial production methods, and an efficient construction process, it will be possible to start using a bridge only two weeks after construction starts on some sites, an improvement of six to eight months over many traditional bridges.

Bridge Deficiencies

The age, condition, and capacity of existing bridges are a major concern. When the I-35 West bridge that crossed the Mississippi River in Minneapolis, Minnesota, collapsed on August 1, 2007, it shocked the nation. Thirteen people died, 145 were injured, and the rest of the nation realized that all of the discussions about the poor condition of the nation's transportation infrastructure were true.

Most bridges are inspected every twenty-four months and receive ratings based on the condition of their various components. Deficient bridges are referred to as being "structurally deficient" or "functionally obsolete." Structural deficiencies are characterized by deteriorated conditions of significant bridge elements and reduced load-carrying capacity. A bridge is considered functionally obsolete when it does not meet current design standards. This can be because the volume of traffic carried by the bridge exceeds the level anticipated for the bridge, or it no longer meets updated design standards.

Bridge Recommendations

Recommendations for developing bridges that fit the concept of green roadways include the following:

Use recycled, recyclable, and industrial by-product materials. Use materials that are stronger, more durable, and less toxic to the environment than traditional materials.

Create a visual design unity among all existing and new structures. Coordinate visual aspects of bridges with sound walls, retaining walls, and other highway structures. Create a visual design relationship that includes coordinating materials, patterns, color, and other design elements.

Integrate landscape and aesthetics at the onset of project planning. The initial report on type, size, and location of highway structures should include information regarding landscape and aesthetics elements.

Use simple substructure and support features. Use simple substructure and support features with strong proportional relationships in bridge design and simple geometric shapes to minimize the support profile.

Use visually light bridge rail structures. Consider open rail design of steel rail or concrete barrier and

FIGURE 3-26 Glulam Timber is ideal for creating bridges and waterfront structures. Its ability to absorb impact forces from traffic makes it frequently used for highway and railway bridges. Image courtesy of Dr. Ronald K. Faller, from Online Guide To Bridge Railings (guides.roadsafellc.com/bridgeRailGuide/index.php).

FIGURE 3-27 The ornamental railing on the Legacy Parkway in Utah is designed to enhance the parkway's surrounding landscape. Photograph by Sharen Hauri, courtesy of National Scenic Byways Online (www.byways.org).

FIGURE 3-28 Open-spandrel bridges, like the Van Duzen River Bridge in Humboldt County, California, are designed to leave more areas of the rail open. Image courtesy of Dr. Ronald K. Faller, from Online Guide To Bridge Railings (guides.roadsafellc.com/bridgeRailGuide/index.php).

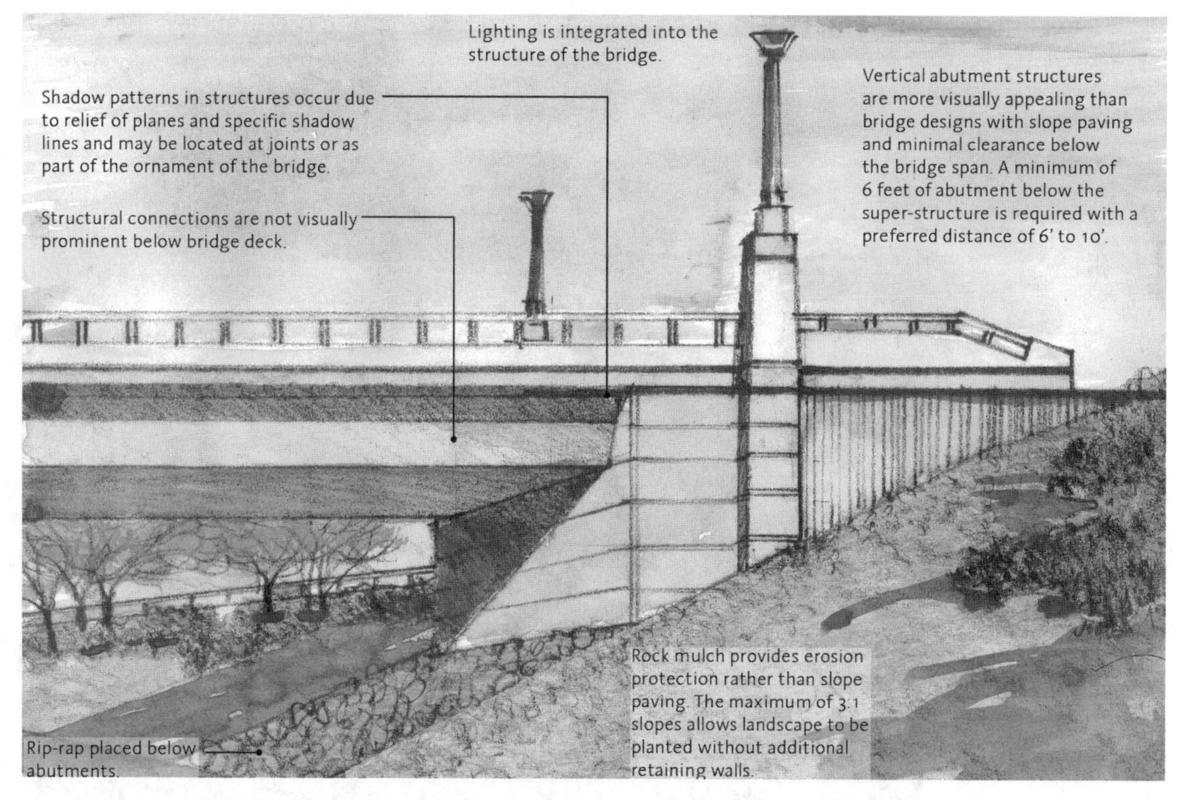

FIGURE 3-29 Nevada's Landscape and Aesthetics Plan developed the idea that bridge abutments and barrier rails should be designed so that they visually fit with other parts of the bridge. Image courtesy of Design Workshop.

steel to create a more refined bridge with a lighter-appearing span. Maintain scenic views and views to the surrounding landscape where possible.

Consider fill embankments and approach rails as part of the bridge design. Consider fill embankments and approach rails in concert with the abutment, bridge barrier rail, and superstructure. Materials, height, and attachment details should be carefully considered when connecting guardrails to the bridge.

Use landscape or rock mulch to stabilize embankments. Contour grade embankments and use landscape planting to maintain embankment where possible. Use retaining walls to establish suitable flat landscape areas where the right-of-way is narrow. Ensure that mulch materials match bridge structure color and the surrounding landscape.

Select vandalism-resistant finishes. Finish type, color, and surface patterns are important design elements in coordinating the structure with the surrounding landscape. Select bridge finishes of appropriate color and vandalism resistance.

Include sidewalks and bicycle accommodation to help improve connectivity.

Establish a schedule that limits heavy construction for bridges during wet times or during breeding season. These precautions will help minimize environmental impacts.

Historically Significant Bridges

Historic bridges are integral elements of the transportation system. Historic bridge research has focused largely on four areas: historic context development, historic bridge inventories, management and preservation plans, and rehabilitation practices. Older bridges have potential historical significance, which adds to their value as community assets, but also complicates the process of rehabilitation or reconstruction of structurally deficient or functionally obsolete

FIGURE 3-30 The BP Pedestrian Bridge in Chicago is a 925-foot bridge that connects Millennium Park to Daley Bicentennial Plaza. Designed by Frank Gehry, the bridge has won awards for its use of sheet metal and is known for its style and aesthetics. Image courtesy of Wikipedia Commons.

Figure 3-31 The strong, curvilinear lines and sleek sheet metal construction epitomize Chicago's emerging new architectural style. Image courtesy Wikipedia Commons.

bridges. State and federal statutes recognize the importance of preserving significant elements of our cultural and engineering heritage. Historically significant bridges are listed or eligible to be listed in the National Register of Historic Places. Historic bridges should be protected wherever possible.

Utilities

Utility easements often run parallel to roads, either located within the right-of-way of a road, or in an easement running parallel to the road. The FHWA has

Figure 3-32 The Bronx River Parkway Reservation is the nation's first public parkway designed exclusively for automobile use. The Bronx River Parkway Reservation Conservancy is dedicated to conserving and enhancing the reservation's natural beauty. Image courtesy of Library of Congress-HABS/HAER.

determined that the use of highway rights-of-way to accommodate public utility facilities is in the public interest, so this is common practice. One problem with this approach is that utilities within the right-of-way can limit future roadway expansion or other uses of the right-of-way.

Overhead utilities, which are strung from one pole to another, typically include electric, telephone, and cable television. In urban and suburban areas, electric boxes, manholes, fire hydrants, lighting, and other fixtures may also be located within or near the right-of-way. One problem with overhead utilities is that motor vehicle collisions with utility poles account for approximately 10 percent of all fixed-object fatal crashes in the United States annually. Utility poles also have a negative effect on the aesthetics of a roadway.

For new construction in urban areas, electric, telephone, and other telecommunication lines are now often placed underground. Power and telephone lines are not buried more often, especially in suburban and rural areas, because this process is very expensive.

Underground utilities may require setbacks for the purpose of access and maintenance.

From a landscape preservation perspective, it is desirable to consolidate, bury, and hide as many of the utilities as possible because they are visual distractions. When this approach is not feasible, utilities and utility zones must be recognized as a landscape design problem that can be mitigated to some extent.

Some states are looking at other ways to use utility poles. New Jersey's biggest utility is adding solar panels to more than two hundred thousand utility poles that are located within road rights-of-way.

Utilities along corridors in areas with important scenic, cultural, or historic resources can be a major distraction. Efforts should be made to relocate the utilities, put them underground, or at least consolidate them to reduce their visual impact.

Pedestrian Facilities

Sidewalks

One key focus of green roadways should be to encourage walkability and improve pedestrian connectivity. In urban areas, sidewalks are often located adjacent to the edge of a road, with a curb providing some separation from the travel lane and the sidewalk. Pedestrian walkways should be provided on both sides of the roadway in all urban and suburban areas. Multiuse trails are typically designed and designated for walking, hiking, bicycling, and other nonmotorized uses such as roller skating and rollerblading. Sidewalks are typically constructed of concrete or some type of paver. Generally, the minimum width for sidewalks is five feet, ten feet for bikeways, and twelve to fourteen feet for shared pathways.

FHWA's *Designing Sidewalks and Trails for Access, Part II: Best Practice Design Guide* is a good source for walkway and trail design.

The Americans with Disabilities Act requires that all sidewalks and trails meet accessibility standards. Grades for pathways used by pedestrians cannot exceed 5 percent unless treated as a ramp, with a maximum slope of 8.33 percent for short stretches of trail.

Along rural roads, sidewalks may not be feasible due to lack of development or destinations, cost, environmental impacts, or other considerations. In these areas, roadway shoulders are often paved and are wide enough to accommodate pedestrians as well as bicyclists.

Planting strips can be used to separate a sidewalk from an adjacent road in both urban and suburban

FIGURE 3-33 At the Watersound development in Florida, emphasis is on walkability. Bridges and walkways are constructed at a pedestrian scale. Image courtesy of EDAW.

areas. In many older, established neighborhoods, these planting strips are wide enough for mature street trees that help define the overall character of the neighborhood.

Where there is sufficient public space adjacent to a road, the creation of a linear park that includes trails, trailheads, seating opportunities, and other amenities is one way to add public open space and help integrate the road into the community.

Trails

Trails are often located outside the roadway clear zone to ensure that pedestrians are protected from motorized vehicles. This improves connectivity while minimizing potential conflicts between pedestrians and motorists.

Although most new walks and trails are typically paved, unpaved trails can also address other recreational and transportation needs. Paved trails are popular for pedestrians, skaters, and cyclists; unpaved trails are often popular with mountain bike riders and cross-country runners.

The National Park Service's definition of a sustainable trail is one that

- supports current and future use with minimal impact to the area's natural systems
- produces negligible soil loss or movement while allowing vegetation to inhabit the area
- allows for pruning or removal of certain plants as necessary for proper trail construction and maintenance
- does not adversely affect the area's wildlife
- accommodates existing use while allowing only appropriate future use
- requires little rerouting and minimal trail maintenance (National Park Service, 1991)

Trailheads are located at key points along sidewalks and trails and serve as access points. Minor trailheads

Figure 3-34 A family walks the edge of the Missouri River on the Washburn Discovery Trail in Washburn, North Dakota. Image courtesy of National Scenic Byways Online (www.byways.org).

FIGURE 3-35 The Freedom Trail is a popular multiuse trail in Atlanta, Georgia. Many people have used the trail for walking, running, and biking since it was completed in 2000. Image courtesy of EDAW.

typically occur where trails connect with sidewalks, and they could include signage, benches, and other community enhancement elements. Major trailheads often occur near parking areas, intersections of major trails, and other key locations where there is expected to be frequent traffic. These major trailheads could include kiosks with signage, covered shelters with seating, drinking fountains, bike racks, benches, ornamental plantings, and other community enhancement elements.

FIGURE 3-36 The trailhead for Sherman Pass Scenic Byway in Washington State gives the hiker information about the journey ahead. Image courtesy of www.pedbikeimages.org / Dan Burden.

Biking Facilities

Accommodating bicycles and other alternative forms of transportation is an essential part of green roadways. Every new transportation project should provide access for bicycles, either along the roadway or via a separate trail. The more we can park our cars, the easier it will be to manage a sustainable transportation system for the future.

As the National Association of City Transportation Officials will tell you, bicycling is good for cities. Safe, comfortable, convenient bicycling facilities provide a cost-effective way for American municipalities to improve mobility, livability, and public health while reducing traffic congestion and carbon dioxide emissions. FHWA predicts that increasing bicycling from 1 percent to 1.5 percent of all trips in the United States would save 462 million gallons of gasoline each year.

A bicycle lane (four to six feet in width) can, with uninterrupted flow, carry a volume of around two thousand bicycles per hour in one direction. A conventional bike lane is located adjacent to motor vehicle travel lanes and flows in the same direction as motor vehicle traffic. According to FHWA, there are five different types of bicycle facilities:

Shared lane–a "standard width" travel lane that both bicycles and motor vehicles share.

Wide outside lane–an outside travel lane with a width of at least fourteen feet to accommodate both bicyclists and nonmotorized vehicles.

Bicycle lane–a portion of the roadway designated by striping, signing, and/or pavement markings for preferential or exclusive use by bicycles and/or other nonmotorized vehicles.

Shoulder–a paved portion of the roadway to the right of the travel way designed to serve bicyclists, pedestrians, and others.

Multiuse path–a facility that is physically separated from the roadway and intended for use by bicyclists, pedestrians, and others. (AASHTO, 2004)

A successful bikeway needs to address bike lanes, intersections, signals, and signage.

The U.S. Bicycle Route System was established in 1982 by AASHTO, which established a new task force in 2003 to study expansion of the system. The following year, the Adventure Cycling Association, a member of the task force, prepared a report called *Developing a National Bicycle Route Corridor Plan* for AASHTO.

FIGURE 3-37 Bike lanes are a common part of all new roads constructed in Baldwin Park, California. Image courtesy of EDAW.

Figure 3-38 In Baldwin Park, California, the combination of bike lanes, landscaped medians, grass edges, and sidewalk are part of the overall community character. Image courtesy of www.pedbikeimages.org / Dan Burden.

The National Association of City Transportation Officials has produced the *Urban Bikeway Design Guide* in an effort to provide cities with state-of-the-practice solutions to help create complete streets for bicyclists. The guide offers assistance for cities seeking to improve bicycle transportation where difficult rights-of-way present unique challenges.

Bicycle racks are an integral part of accommodating bicycle transportation. They help encourage cycling, provide ways to store bikes safely, and discourage users from locking bikes to railings, street trees, and other furnishings. Facilities such as Bikestations in Washington, D.C., which provide parking for more than one hundred bicycles, also include a changing room, lockers, bike rental, bike repair, and retail sales.

Signage

Signage is important to help motorists, pedestrians, and cyclists understand critical information related to a transportation corridor. Signs also have a major impact on the aesthetic character of the roadway.

The federal *Manual on Uniform Traffic Control Devices* defines the standards for all traffic control devices, including road markings, highway signs, and traffic signals. These standards help ensure uniformity and consistency nationwide, which helps reduce crashes and congestion, and improves the efficiency of highways and roadways. Although the manual's standards are fairly rigid, they are not an obstacle when it comes to green roadways. It is still possible to incorporate custom interpretive signage or way-finding signage, such as those you may find along scenic roadways.

Among the many types of signs used along roadways are the following:

Regulatory signs–These include stop and yield signs, speed regulations, turn and lane use, directional, parking regulations, traffic signals, railroad, weight limit, and seat belt signs, among others.

Warning signs–These include turn and curve, intersection, merge and lane transition, width restrictions, divided highway, hill, clearance, advisory speed, work zone, and slow traffic, among others.

FIGURE 3-39 The colorful entrance sign to the River Road Scenic Byway in Huron-Manistee National Forest, Michigan, welcomes visitors to the forest. Image courtesy of National Scenic Byways Online (www.byways.org).

FIGURE 3-40 Directional signage in Portland, Oregon, helps orient visitors. Image courtesy of www.pedbikeimages.org / Heather Bowden.

Marker signs–Marker signs include route markers, junction signs, alternate route signs, turns, and directional signs.

Guide and informational signs–These include destination and distance, expressway and freeway, work zone information, general information, and cultural signs.

Tourism and recreational signs–These include general information signs, traveler services, accommodation services, and recreation signs. Tourist-oriented directional signage is intended to offer a cohesive, understated manner for businesses to advertise. These signs cover cultural, historical, recreational, educational, or entertaining activities, or unique commercial activities whose major portion of income or visitors is derived from tourists. Each business pays a fee to have its advertisement on the tourist-oriented sign.

In addition, there are also temporary traffic signs, school signs, bicycle signs, and emergency and incident management signs.

All sign supports on highways within the clear zone must either be of a breakaway type meeting the crashworthiness criteria of National Cooperative Highway Research Program Report 350—*Recommended Procedures for the Safety Performance Evaluation of Highway Features*, or be protected by guardrail, barrier, or an energy-absorbing system meeting this report's criteria.

Wood and steel are the two primary materials used for small sign supports. Larger sign supports, such as cantilever structures or sign bridges, are usually made of steel. Whenever reasonably feasible, existing structures such as overpasses should be used for support of overhead signs. This reduces cost and improves safety by minimizing the number of separate structures needed.

Outdoor Advertising

Outdoor advertising, specifically billboards, provides businesses, community groups, and other organizations opportunities to inform travelers along the roadway about the various establishments and available services. However, billboards affect the visual quality of the highway because they obstruct views of scenic features and the natural landscape. As a result, some community groups are committed to restricting new and removing existing billboards from areas near and within their communities.

The Highway Beautification Act, which was passed in 1965, was intended to control billboards along highways. However, in the almost fifty years since the passage of the act, few nonconforming billboards have been removed and many more have been constructed due to exclusions in the law. Enforcement is difficult because of Section G of the law, which requires cities and counties to pay just compensation to owners for billboard removal. A second limitation in the act is the allowance of billboards to be constructed in areas zoned commercial and industrial, as well as in unzoned areas with commercial or industrial uses.

Cities and counties have the ability to regulate the location and to a limited degree the type of billboard erected within their jurisdiction. The development of design standards that address height, size, color, and context in which the billboards are located is a valuable method of directing outdoor advertising. The visual impact of billboards in the rural landscape is much greater than the impact generated by billboards in an urbanized location. Standards, such as setbacks, not allowing billboards along portions of a road, and size limitations, can be established to help control billboards.

Place Name Signs

A place name and point-of-interest sign program that is distinctive to a particular corridor or region will better connect people to places. The basic idea behind this program is to use signage to let both locals and visitors gain a better understanding of the unique features around them.

Audio Programming

One way to reduce the amount of signage along roadway corridors is to use audio/multimedia programming that provides motorists with information about tourism as well as other information that could be tied into a state's sign program. This program could be implemented via broadcast radio, CD, DVD, wireless Internet hot spots, satellite transmission, or other media. Travelers could access from their car additional information about cultural and natural resources, tourist opportunities, and services along the corridor.

Lighting

Lighting along roadways is often necessary for safety and security. Street and highway lighting in the United States currently accounts for 2 percent of overall electrical demand. Today's road standards for lighting emphasize aesthetics, architectural qualities, and efficient fixtures that minimize light pollution. A well-lighted structure or design element along the highway can do a lot to improve the overall perception of a highway corridor. Lighting in rural areas should be limited by dimming it at night and choosing a low lighting classification.

Recent studies have found that light-emitting diode (LED) technology is becoming competitive for streetlight applications with the commonly employed high-intensity discharge light sources such as high-pressure sodium and metal halide. The expectation is that LED street lighting technology will not only provide more efficient light distribution and increased uniformity, but will also save energy and reduce maintenance costs.

FIGURE 3-41 Along US 93, which runs through the Flathead Indian Reservation in northwestern Montana, place name signs that identify key cultural and natural resources are given in Salish and Kootenai as well as in English. Image courtesy of Jones & Jones.

A number of DOTs and municipalities have investigated or piloted shifts to LED- and solar-powered lighting for signs. Using energy-saving LED lightbulbs would eliminate nine million tons of carbon dioxide emissions per year. Washington DOT has already converted 90 percent of the signals they maintain to LEDs. California DOT is also switching from incandescent to LED lights. Los Angeles has committed to replacing 140,000 existing street lamps with LEDs over a five-year period. The City of San Jose intends to replace all 62,000 streetlights throughout the city before 2022 as part of their "Green Vision" program.

Seattle City Light has a street lighting system of nearly 84,000 street and area lights that use predominantly high-pressure sodium light sources. Because of the potential benefits of installing LED luminaires as a replacement for these lights, Seattle City Light launched a study to evaluate LED luminaires for photometric performance, energy efficiency, economic performance, and the impact of the new lights on the city's streetlight system.

New research has shown that the coordinated use of lighting and roadside vegetation can be an efficient approach. An ecoluminance approach uses lower power luminaires than typically used for roadway lighting to illuminate and provide reflected luminance of objects such as vegetation. The ecoluminance approach can result in lower energy use and have a lower initial cost, operation cost, and maintenance cost than conventional lighting.

Along transportation corridors, lighting fixtures

can be used that minimize light pollution and provide even light dispersion. As a general rule, very tall light posts, called high-mast lighting, should be avoided. Excessive high-mast lighting can create light pollution along a corridor and excessive height masts can affect the view of surrounding vistas.

Illinois DOT is using retroreflective sign material on the front of roadway signs to nearly double their effective brightness. This means the letters on a sign light up, much like the eyes of a cat in the dark. The lunar-resonant streetlamp is a light fixture that is designed to measure the strength of the moonlight and adjust its brightness correspondingly. Philips has developed a fixture called the "Light Blossom" that uses solar panels to harness the sun, and a lamp post that moves to harness power from wind. The Light Blossom is designed to change to match existing conditions. During the day, the lamp's petals open to reveal solar panels, while the head can rotate to follow the sun. During windy conditions, the petals contract to a half-open position and turn into a rotating wind turbine. At night, the panels close and the fixture is used to light the surrounding streets (Freemark, 2010).

Parking

With all of the cars on the roads, we have to have some place to park them. As a result, parking lots have become a dominant feature in the landscape, especially in urban and suburban areas. In a typical U.S. city, paved surfaces make up 30–45 percent of the total area, and approximately 15 percent of that surface area consists of parking areas. Traditionally, most parking lots have been constructed of concrete or asphalt, which has led to increased stormwater runoff and reduced open space and affected natural resources.

Stormwater runoff results in large volumes of polluted runoff entering surface water and groundwater resources, negatively affecting water quality. Taking a green approach to parking lots is one of the best ways to address water quality and pollution issues.

The U.S. Environmental Protection Agency has produced the Green Parking Lot Resource Guide, which is a good source of information on creating more environmentally friendly parking lots. Some recommendations in this guide, and others, are:

FIGURE 3-42 A bioswale, like this one along 12th Avenue in Portland, Oregon, is used to reduce runoff volumes by slowing down water through vegetation. Image courtesy of Kevin Robert Perry, City of Portland.

FIGURE 3-43 Using a green approach to parking areas, such as this one in Shekou, China, allows rainwater to drain through the pavement, thus reducing pollution. Image courtesy of EDAW.

Reduce minimum parking requirements–One problem is that too often in the past, parking lots were overdesigned to accommodate the largest number of customers. A good starting point for creating a greener world is to minimize the number and size of surface parking lots and explore other alternatives for parking.

Establish maximum limits on parking–A maximum number of parking spaces should be allowed within a given area based upon square footage of buildable space.

Reduce stall dimensions–The size of parking spaces should be changed because of the shift to smaller vehicles. In the old days, parking spaces would be ten feet by twenty feet, but now they are reduced to nine feet by eighteen feet, or sometimes as narrow as eight feet or 8.5 feet. This smaller size works as automobiles get smaller too.

Provide on-street parking–On-street parking provides a buffer to pedestrians from traffic and has been found to decrease traffic speeds. On-street parking provides a buffer between motor vehicle traffic and pedestrians on the sidewalk, but should be considered only on secondary streets with speed limits less than thirty-five miles per hour. On-street parking can also influence the traffic flow along roadways, sometimes resulting in reduced speeds, reduced capacity, and increased conflicts for both bicycle and motor vehicle traffic.

Improve the visual quality of parking lots–Adding planting medians that are large enough to support shade trees enhances visual quality and reduces the heat impact of paving areas. The medians can also be used for bioswales that help manage stormwater runoff.

Use alternative parking approaches–Many cities are encouraging alternative parking strategies, such as reduced parking ratios or having businesses share parking, to encourage pedestrian activity in key areas, such as downtowns and near transit.

Reduce visual impacts–Providing buffers between roads and parking areas can greatly reduce the visual impact of parking lots.

Incorporate stormwater best management practices into parking lot design–Stormwater best management practices include structural controls and bioengineering techniques designed to facilitate natural water cycling processes by capturing, filtering, infiltrating, and/or storing stormwater.

Use pervious pavement–Types of permeable and semipermeable alternative pavers include gravel, cobble, concrete, wood mulch, brick, open-jointed pavers filled with turf or aggregate, turf blocks, natural stone, and pervious concrete.

Use in-lieu parking fees–Instead of requiring each developer to provide parking, there could be an option to pay fees to a municipality that can be used to build a central parking structure that can help meet the needs of many.

One alternative approach to surface parking is Green Park DIA, a 4,200-stall parking facility servicing Denver International Airport. The $32-million facility is being built to meet LEED Gold standards. It uses solar, wind, and thermal energy and harnessed methane from a neighboring landfill that is converted to natural gas. The parking lot is constructed of porous pavement, and native plants are used to create a more environmentally friendly site. Alternative-fuel shuttles are used to take customers to the airport terminal and back.

Facilities

Welcome centers, rest areas, scenic overlooks, and other similar transportation facilities can be an important part of a roadway. They have a significant impact on the driving experience, address much-needed utilitarian purposes, and provide an opportunity to highlight key cultural, natural, and visual resources along the roadway.

Welcome Centers

Welcome centers can be promoted as important civic facilities. They can have a strong visual presence or visually blend into their surroundings, depending on the character of the road. Welcome centers at state borders should convey the identity of the place and make the entry into the state a notable and memorable experience. The center should be visually appealing both day and night and connect travelers with the natural landscape and scenic views.

Rest Areas

Basic rest areas can be located throughout the state and offer site-specific interpretive information. They have limited restroom facilities, which may or may not include running water, depending on availability. Typically, these rest areas are located to take advantage of scenic views, unique historical, cultural, or environmental features, and to provide travelers resting places en route.

Complete rest areas are typically located at sixty-mile intervals throughout the state and are typically situated outside of developed areas. A typical rest area requires approximately twenty-five acres of land within the highway right-of-way to accommodate the site functions. Rest areas in the middle of a highway are not recommended because they result in left-hand exits and entrances to the highway.

Roadside Pull-offs

Roadside pull-offs provide facilities for drivers to exit the highway for a brief period. Facilities that respond to the landscape character and provide minimal parking accommodate the abbreviated stay.

Viewpoints and Points of Interest

Viewpoints and points of interest present opportunities to view unique vistas, special natural resources, historical features, or cultural landmarks. Interpretive elements are integrated into the site design, and place name signage and travel information elements are provided to establish the relationship between highway and place. Typically, the length of stay is short and parking is limited.

Figure 3-44 Located at the base of the National Children's Forest Land is the Children's Information Center. Newly remodeled in 2010, this unique nature center is the only one in the country operated by youth volunteers. Photo by Martha Alejandre, courtesy of National Scenic Byways Online (www.byways.org).

Figure 3-45 This rest area in Arizona provides travelers with local information, maps, restroom facilities, and a scenic view. Image courtesy of FHWA.

Figure 3-46 Part of the National Wild and Scenic Rivers System is the fully paved Cascade Stream Watch Trail. Along the path you will find a handful of informational maps and signs that cover a wide array of topics. Image courtesy of Wolftree, Inc.

Park and Ride Lots

Park and ride lots serve as collection points for motorists who seek to park their car and use some kind of mass transit. This may include buses, trains, light-rail, or ride sharing. Other activities potentially supported at a park and ride facility include pedestrian, bicycle, paratransit, intercity bus transit, airport service, and intercity and commuter rail. The design of a park and ride facility should consider expected parking demand, parking facility design standards, and circulation patterns. These can also be associated with transit centers that offer a higher degree of travel services, route choices, and mode choices.

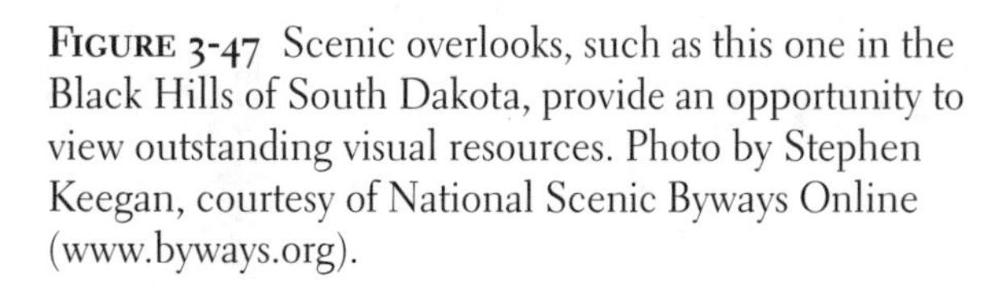

Figure 3-47 Scenic overlooks, such as this one in the Black Hills of South Dakota, provide an opportunity to view outstanding visual resources. Photo by Stephen Keegan, courtesy of National Scenic Byways Online (www.byways.org).

Plants

Plants are a major part of green roadways. They help determine the overall visual character of a landscape, and are an important part of the overall health of an ecosystem.

Every green roadway should emphasize protection of existing vegetation and planting native materials to help mitigate environmental impacts, enhance visual quality, and create a healthier environment. In landscape design and planning, understanding the contextual plant communities surrounding a site is of paramount importance in providing an ecological basis for sustainable approaches to design. Plant communities consist of fairly predictable associations of individual plant species that tend to occupy a particular ecological niche in the landscape.

According to the Texas DOT, the use of the roadside for specialized environmental goals should be carefully considered to be sure that the safety, sustainability, and life-cycle costs of the project meet department goals and resources. One approach is to restore a site to the topographic shape, hydrologic function, and plant community that existed before disturbance by people. This practice is expensive and requires detailed knowledge and constant management. A more manageable approach is to develop plant communities for use as habitat by birds, mammals, reptiles, or insects. Habitat creation involves providing one or all of cover, food, or water to a targeted species and requires detailed planning and development funding. In addition, naturalization seeks to promote or reintroduce native plants to minimize maintenance or improve the aesthetics of the roadside. This usually involves the seeding or planting of desirable plants and periodic management to assist in their survival, or it may focus on preserving threatened or endangered species.

A key for every transportation project is to ensure that the plant palette selected for the site complements existing vegetation in the surrounding landscape. Wise plant selection will create sustainable plant communities and transition areas, benefiting

FIGURE 3-48 More than 12,000 acres of Iowa's roadsides have been seeded with native plants that provide low-maintenance weed control and enhance the beauty of the landscape. Image courtesy of Iowa Department of Transportation.

FIGURE 3-49 Iowa's Integrated Roadside Vegetation Management program was developed to use native vegetation to produce a more cost-effective, environmentally friendly way of managing roadside weed and erosion control. Image courtesy of Iowa Department of Transportation.

human and ecological systems, and will reduce maintenance in the long run. Understanding the site context leads to selection of the most appropriate plants.

All plant materials should be selected to provide a valuable landscape amenity that is both attractive and meets sustainability goals. Canopy trees, understory trees, shrubs, groundcovers, perennials, and other plant material should be used to add visual interest, help create a healthier environment, and complement existing vegetation. Native plants can offer unique richness and aesthetic interest in landscapes often overlooked by typical landscape development. Ground treatment should be consistent in size, texture, color, and aggregate mix with the surrounding landscape.

Some of the recommendations for implementing plantings into a green roadway are as follows:

Preserve healthy, mature trees and/or vegetation within the right-of-way. Mature vegetation is an integral part of community life and an important public resource that enhances the quality of life.

Take an ecosystem approach to revegetation along a corridor. Many transportation projects extend for miles and they can traverse many different landscape types. Design should be sensitive to changing conditions.

Integrate landscape and aesthetics at the onset of planning, design, and engineering of all highway

FIGURE 3-50 Planting native vegetation in mass can have a major visual impact and is consistent with how the vegetation would occur naturally. Image acquired from depositphotos.com.

projects. Landscape and aesthetics should not be an afterthought to a highway project.

Salvage native plants and topsoil prior to construction. The species to be salvaged depend on location, soils, and analysis of plant value, including the potential survival rate.

Collect native seed as part of a specific transportation project. Initiate a process for native seed collection at the start of each project.

Remove invasive species. Invasive species can deteriorate economic and environmental quality and cause harm to human health. Invasive species are common along roadsides because they survive and thrive in harsh climates and poor soils.

Incorporate wildflowers into corridors. Natural sites along the road offer an ideal opportunity for cultivating wildflowers. In addition to improved aesthetics, native wildflowers can also reduce maintenance requirements; reduce roadside fire hazards; reduce use of herbicides; improve erosion control; and improve the relationship between the highway corridor and the regional character of the landscape.

Use mulch. Mulch is especially useful in establishing planting beds, managing weeds, conserving moisture, and amending the soil as a long-term conditioner. Organic mulch offers the most sustainable option for green roadways.

Use street trees. Vegetation and individual street trees play an important role along streetscape corridors.

Art

Transportation facilities can provide a great opportunity to showcase public art elements that provide aesthetic and cultural benefits to a community. These benefits can potentially result in positive economic development, increased tourism, and a greater sense of place. Public art can be many things, including sculptures, obelisks, columns, castings, murals, paving patterns, landforms, fountains, or other site elements. These artistic features may be freestanding or placed on roadway structural features such as noise walls, retaining walls, sidewalks, bridges, bike paths or other approved engineered structures.

Art programs can enhance the quality of public spaces, reflect local culture, and provide a venue for community engagement in project planning and design decisions. One of the major benefits of

FIGURE 3-51 Along Abbot Drive in Omaha, Nebraska, large columns on either side of the road visually function as a gateway and landmark. Image courtesy of EDAW.

integrating art into transportation projects is that it provides an opportunity for stakeholders to have a say in the final look of an interchange, crossing, transit station, or other transportation facility.

One program that funds numerous arts projects nationwide is the Transportation Enhancement program, which is funded through the U.S. Department of Transportation and administered by state transportation agencies in partnership with local arts agencies.

Some recommendations for integrating public art into transportation projects include:

Create regionally appropriate, meaningful art. For the roadway user, an art-scape enhances the travel experience and the impression of place. Transportation art should be authentic and should evoke clear meaning and purpose that relates to the surrounding place, the unique culture and environment of the area, and the travel experience.

Complement highway structures. Enhance bridges, pedestrian structures, sound walls and retaining walls with appropriate motifs and consider sculptural ornamentation, decoration, and landmark features. Patterns imprinted on a highway structure should be designed with an artistic composition. While complementing other highway structures in form and color, patterns should offer a level of complexity and interest that responds to the unique experience of the place and roadway travel. Artwork should be of a scale appropriate to highway travel speed.

Ensure that artwork expresses excellence of craftsmanship, quality, truthfulness, and originality.

FIGURE 3-52 In 2006, as part of CowParade, a traveling art initiative, more than one hundred decorated life-size fiberglass cows were placed throughout Madison, Wisconsin. The basic idea behind the exhibit was for Wisconsinites to have fun with the dairy history of the state. Image courtesy of www.pedbikeimages.org / Dan Burden.

Elements of highway art should not be obvious or inauthentic. Avoid the use of ready-made, randomly placed, stand-alone objects or imprints that depict little meaning.

Consider each art piece as part of a larger whole. Highway art can be carefully crafted and the simplest of all elements have a very powerful effect. When planning transportation art, the entire length of each design segment and the corridor should be considered.

Engage local agencies and organizations in the planning process. Artwork can be included as a component of landscape and aesthetic projects or as freestanding art installations. Consider transportation art at the onset of project development. Engage community members, artists, landscape architects, and architects early in the design and development stages of highway projects to ensure an integrated and comprehensive art program (Design Workshop, 2006).

Consider visual impact. Art must conform to the Outdoor Advertising Act and cannot display text, images, flags, religious or political symbols, logos, commercial images, or other elements deemed by the highway department as advertising or sponsorship. Art must not be offensive to motorists, pedestrians, and local communities or create negative public reaction.

Maintain safety. Art must not create safety hazards for motorists and pedestrians. Sculptures and freestanding art must be installed outside the clear recovery zone along the perimeter of the roadway and outside of sight distance areas.

Minimize visual distractions. The installation of art with highly reflective surfaces, colors, moving parts, or flashing or bright lights is strongly discouraged.

Control access. Public access is prohibited on high-volume and high-speed facilities.

Summary

All roadways have some combination of design elements that collectively define the road and its impact on surrounding environments. The real question is how these basic roadway design elements take into consideration cultural, natural, and visual resources while addressing safety and connectivity. In the United States, we are very good at creating roadways that are safe and functional. These successes are based upon roadway design standards for parameters such as horizontal and vertical curves, shoulders, slopes, signage, and other design elements. It is possible to meet traditional engineering standards such as level of service and design speeds while also creating roadways that are attractive, are community assets, and minimize impacts on the environment.

Chapter 4

Design and Planning Process for Green Roadways

After sixty years of a transportation planning approach based on the concept that more paving is better, we may finally be at the cusp of a new era. The American Association of State Highway and Transportation Officials' (AASHTO) *Policy on Design of Urban Highways and Arterial Streets* (2011), commonly known as the "Red Book," advocates a design approach where highways have a "tailor-made design for the unique set of conditions along the segment." This sentiment is at the heart of a green roadways approach.

During the 1990s highway design changed rapidly throughout the United States. Highway designers and builders learned they must be more sensitive to the impact of highways on the environment and communities. "In the past, the transportation engineers would show up at a meeting and tell everyone what we needed to do," says Wes Hughen, a transportation engineer with Tennessee Department of Transportation. "We have changed how we approach projects and we have gotten a lot better at listening to what people want."

A number of different design and planning approaches can be used for transportation projects. These approaches are often parts of a more holistic approach to creating green roadways. One reason there are different design and planning approaches is that so many professions are involved in creating a green road. The key is to integrate these processes to make sure that all perspectives are addressed for each transportation project.

Overview of the Process of Planning and Designing a Green Road

There are many different approaches to designing a road at the site level, but they can all be simplified into four major steps: research, inventory/analysis, synthesis, and implementation. These steps are best applied at specific locations where the emphasis is on implementing a project, rather than broad, planning-level studies.

Research–One of the initial steps in the design process is to determine the focus and extent of a transportation project. For a project to meet its stated goals, it needs to be considered within a broader context. This involves understanding local and regional transportation trends and patterns and anticipated changes that could affect transportation infrastructure. Researching analogous transportation projects may provide insight on possible approaches and solutions for a specific project.

Inventory/Analysis–Effective transportation design and planning begins with a sound base of information on the opportunities and constraints for each transportation project. For every project, designers and planners conduct an initial inventory and analysis of a corridor and adjacent areas. This information serves as the foundation from which decisions are made. The inventory phase includes identifying significant cultural, natural, historic, and aesthetic resources and evaluating them as part of the transportation planning process.

Part of understanding local needs is to listen to residents and other stakeholders. Stakeholder involvement will be paramount in generating enthusiasm and building broad community support for innovative transportation planning.

Synthesis–This phase involves building on the research and inventory/analysis and generating a concept and design that meet the goals of a specific transportation project. The objective of the synthesis process is to develop a master plan that provides clear design guidance for a cohesive series of implementation projects. This generally is the responsibility of designers involved with a project, but stakeholder and client input are also important.

Emphasis is on designs that are sustainable, from both an environmental and an economic standpoint. Public meetings give citizens and stakeholders opportunities to share their thoughts on the master plan. Feedback from the city, stakeholders, and other appropriate participants is incorporated into the design.

Implementation–The final measure of any transportation plan is whether its recommendations are implemented. A plan needs to be achievable and sustainable, balancing community vision and the fiscal responsibility of a transportation department. An action plan and implementation strategy is often developed that notes tasks, responsibilities, and timelines for moving the project forward.

Design guidelines are intended to be a framework by which new site-specific projects along the transportation corridors can be assessed and critiqued, as well as a reference for improving landscape and aesthetics when designing new and retrofit highway projects. Fundamentally, they help ensure that many of the green elements that were agreed upon during the planning process are implemented as intended. They also define the look and feel of a given project to help visually fit with a community.

These guidelines are not meant to be prescriptive, but to allow a range of design solutions to achieve the primary principle at hand, providing recommendations on the types of materials, color, form, texture, style, and other level of detail needed to ensure a consistent visual character along transportation corridors. The focus is on consistency and continuity, not uniformity. The design guidelines cannot anticipate every project proposal that may arise, so each project needs to be reviewed on a case-by-case basis.

One of the most important factors in executing a transportation project that addresses the environmental, cultural, historic, and visual concerns is to assemble a diverse, interdisciplinary team committed to excellence and creative problem solving. The interdisciplinary team helps meet transportation safety and mobility goals by applying creative solutions that meet the specific project needs rather than borrowing solutions from standards manuals and past projects.

The earlier the interdisciplinary team is formed, the better. Including landscape architects, architects, planners, engineers, urban designers, and artists in the transportation planning process typically results in a project that is more innovative and holistic, and is embraced more by local stakeholders. This process may take a little longer than a traditional engineering approach, but the results are well worth the extra time needed.

Environmental Considerations in Planning

For many transportation projects, the planning process actually occurs in two steps. First, the National

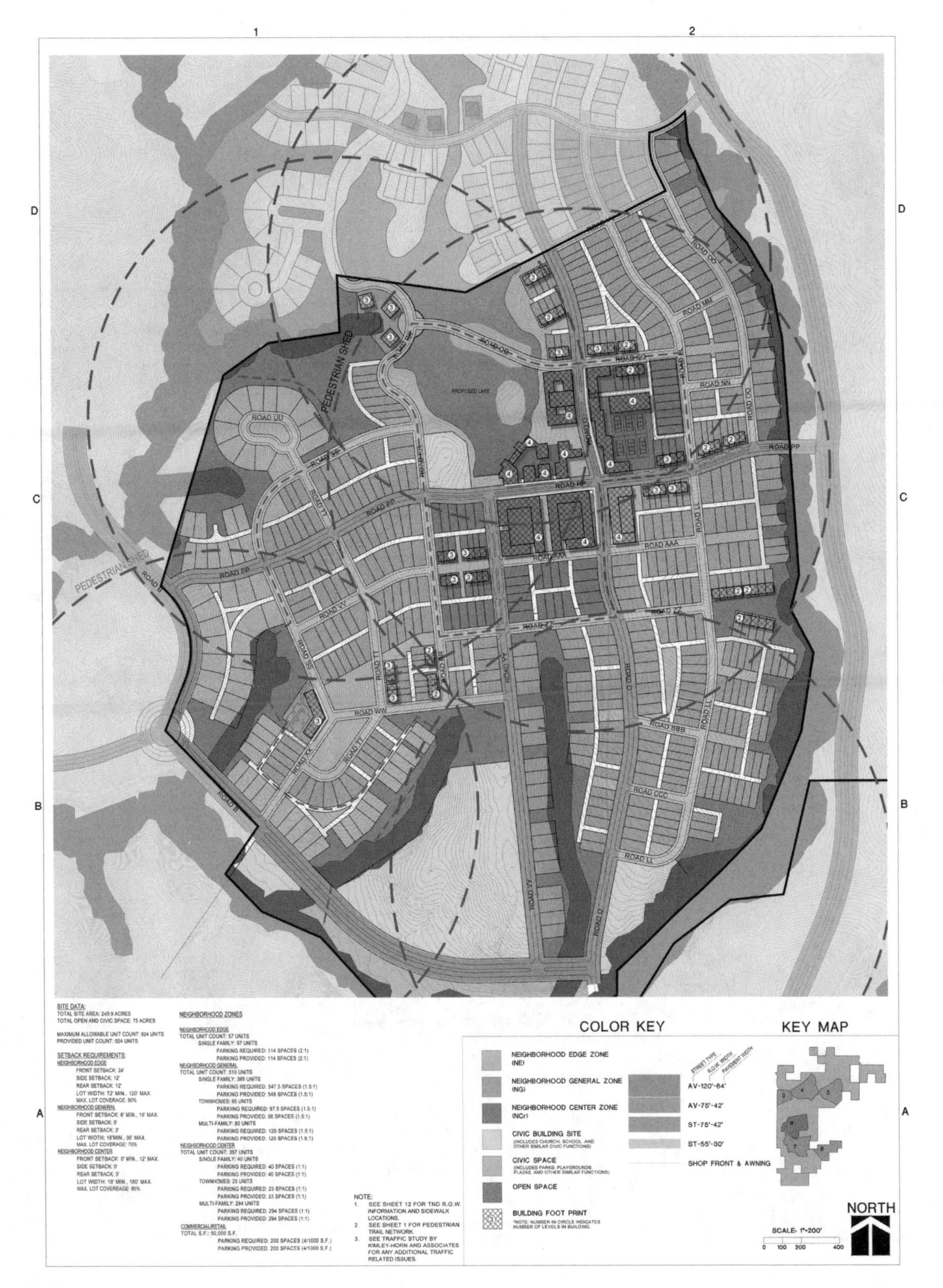

FIGURE 4-1 During the synthesis phase of the design process for A Village in the Forest, a proposed mixed-use development north of Atlanta, Georgia, various land use and transportation patterns were explored to determine which would be most effective while still protecting cultural and natural resources. Image courtesy of EDAW.

Figure 4-2 The Monroe Connector/Bypass is toll road along a twenty-one-mile corridor in central North Carolina. Aesthetics were a major consideration, with the overall theme being based upon a combination of brick and stone surfaces to give the roadway's bridges a classic look. Image courtesy of AECOM.

Figure 4-3 Various concepts were explored for the bridges of the Monroe Connector/Bypass. Major bridges showcased a feature called a freestanding pylon that is reminiscent of a clock tower. Image courtesy of AECOM.

Figure 4-4 The concept of green roads places an emphasis on getting people out of their cars and creating walkable areas that help define the visual quality of a community. Image courtesy of www.pedbikeimages.org / Carl Sundstrom.

Figure 4-5 Simple corten steel pedestrian bridges are an inexpensive way to enhance connectivity to trails and recreation opportunities. Image acquired from depositphotos.com.

Environmental Policy Act process must be completed to ensure that environmental impacts are considered before selecting a preferred alternative. The second step is then to initiate a process that leads to implementation of the preferred alternative. For more detail, see chapter 3. In addition, particular environmental concerns may need to be addressed.

Watershed Impact

A watershed plan is a strategy that provides assessment and management information for a watershed, including the analyses, actions, participants, and resources related to developing and implementing the plan. Typically watershed planning is not specifically integrated with the transportation planning process, but it should be part of a green road strategy when the project interacts with water resources.

Comprehensive water management planning usually is conducted by entities that range from water districts and large multicounty urban areas to state water resources agencies and regional river basin commissions.

Considering Green Space

Conservation is an important part of any planning effort and is one of the highest priorities in helping to ensure that we are able to balance human needs with environmental requirements. Too often, roadway projects are proposed for undeveloped green space because it is easier to build there without disrupting existing neighborhoods and cities. In the past, people would speak loudly to protect their homes, but they were not as adamant about protecting the landscape. As roads have proliferated, citizens have become more aware of landscape fragmentation and are often vocal about protecting the open space that remains, and their concerns should be taken into account.

Transportation Planning

Transportation planning is typically emphasized when the main goal of a road is to address general transportation issues for a neighborhood, city, or region. This approach includes a number of steps:

- Monitor existing conditions..
- Forecast future population and employment growth, including assessing projected land uses in the region and identifying major growth corridors.
- Identify current and projected future transportation problems and needs and analyze, through detailed planning studies, various transportation improvement strategies to address those needs.
- Develop long-range plans and short-range programs of alternative capital improvement and operational strategies for moving people and goods.
- Estimate the impact of recommended future improvements to the transportation system on environmental features, including air quality.
- Develop a financial plan for securing sufficient revenues to cover the costs of implementing strategies.

Once the planning approach is completed, detailed design and construction documents need to be prepared for each transportation project. The intent of these drawings is to provide sufficient information to contractors to construct the road.

Rethinking Roads: Context-Sensitive Design/Context-Sensitive Solutions

The idea that the geometry of pavement is the primary focus of transportation planning is being gradually overtaken by a shift to a more holistic approach that many refer to as context-sensitive design (CSD) or context-sensitive solutions (CSS). Most transportation projects these days strive to be more environmentally friendly, more attractive, and more responsive to cultural concerns without having to sacrifice safety and accessibility.

In 1998, the Maryland Department of Transportation, Maryland State Highway Administration, AASHTO, and Federal Highway Administration (FHWA) held a conference entitled "Thinking Beyond the Pavement." This conference set out the basic concepts for CSD, which was defined as "a collaborative, interdisciplinary approach that involves all stakeholders to develop a transportation facility that fits its physical setting and preserves scenic, aesthetic, historic, and environmental resources, while maintaining safety and mobility" (NCHRP, 2009).

At the workshop, Tom Warne, former executive director of Utah Department of Transportation, said, "In the beginning of the Interstate era, we built the greatest freeway system in the world; but aesthetics and preserving the environment weren't part of that mission. Now we need another transformation. We're here to define a new vision, to change how we do business" (NCHRP, 2009).

In recent years, CSD has been replaced with CSS, which gained popularity because it suggests a wider spectrum of context-sensitive issues that includes the entire process from planning through construction. CSS principles are as follows:

- Use interdisciplinary teams.
- Involve stakeholders.
- Seek broad-based public involvement.
- Use a full range of communication strategies.

FIGURE 4-6 Seaside, Florida, is a planned community that was designed around the idea of creating an old-fashioned beach town that emphasizes walkability and social interaction. Image courtesy of EDAW.

- Achieve consensus on purpose and need.
- Address alternatives and all modes of transportation.
- Consider a safe facility for users and the community.
- Maintain environmental harmony.
- Address community and social issues.
- Address aesthetic treatments and enhancements.
- Use a full range of design choices.
- Document project decisions.
- Track and meet all commitments.
- Use agency resources effectively.
- Create a lasting value for the community. (NCHRP, 2009)

CSS has been around long enough that it has slowly become part of the transportation culture. As a result, we are now seeing the development of more green roads than in the past. In 2003, FHWA established a goal to incorporate CSS into planning and project development in all fifty states by 2007. Today, every state department of transportation (DOT) uses CSS to some degree. There is a consensus that incorporating CSS principles early in the transportation planning process is likely to save money and reduce project delays.

Another potential benefit of CSS is that the process can lead to a better working relationship between transportation agencies and stakeholders. CSS requires early contact with the public and other stakeholders to get input. See chapter 5 for the benefits of stakeholder input.

FHWA launched www.ContextSensitiveSolutions.org in 2004. The clearinghouse is an Internet-based national resource center that contains built examples of CSS projects, case studies, cutting-edge research, information, and policy documents. It addresses a broad range of issues, including design standards, liability, stakeholder involvement, and new techniques in transportation problem solving, and is a good resource for creating green roadways.

Planning to Reduce Risks Associated with Green Roadways

Safety is a top priority for all transportation projects, and looking for ways to reduce risks is an integral part of a green roadway planning process.

According to the National Highway Traffic Safety Administration, deaths and injuries resulting from motor vehicle crashes are the leading cause of death for Americans of every age from three to thirty-four. A statistical projection of traffic fatalities for the first nine months of 2011 shows that an estimated 24,050 people died in motor vehicle traffic crashes. This represents a decline of about 1.6 percent as compared to the estimated 24,437 fatalities that occurred in the first nine months of 2010. In addition, motor vehicle crashes cost approximately $230 billion per year in the United States.

We certainly want an environmentally friendly, sustainable road that looks good and includes amenities such as trails, open space, and stormwater management, but we can do all of that while also ensuring that the road is safe. One way to ensure that a transportation project is safe is to manage potential risks.

Risk management is the process of identifying, evaluating, prioritizing, and mitigating risks associated with a project. Part of this process is assessing the probability of certain risks occurring. Green roadways should not have more risks than any other road.

Some have been hesitant to adopt CSS and other green planning approaches because of the concern that doing so would result in accepting a higher risk. Research has shown, though, that risks are actually lower with CSS and other similar approaches. A major reason is that drivers slow down because the setting is more attractive, and because traffic-calming measures help reduce potential conflicts. Flexible design and planning decisions can enjoy protection from claims of negligence as long as the designers can show that, in fact, they exercised this discretion by carefully evaluating alternatives and weighing the important trade-offs.

An agency's management structure and project development processes are both important for good risk management. One of the best ways to reduce risks is to have a process in place that ensures decisions are made in a pragmatic fashion. Documenting the decision-making process helps reduce the likelihood of a legal challenge. Ways to integrate safety considerations into the transportation planning process include Strategic Highway Safety Plans, Statewide Transportation Improvement Programs, and Transportation Improvement Programs developed by the state DOTs and metropolitan planning organizations.

The Safe, Accountable, Flexible, Efficient Transportation Equity Act: A Legacy for Users (2005) established the Highway Safety Improvement Program, which seeks to reduce traffic fatalities and serious injuries on all public roads through the implementation of infrastructure-related highway safety improvements.

In addition, some of the approaches in the following sections related to land use planning, such as road diets and complete streets, emphasize improved safety for pedestrians, cyclists, and drivers.

Land Use Planning, Smart Growth, Complete Streets, and Transportation Infrastructure

Land use planning focuses on the physical layout of communities, and is closely linked to transportation planning. The implementation of land use policies can help communities and regions balance the demands of growth with the health of environmental resources. These policies determine how streets, pipes, and water lines are laid out and where roads and parks are located. They also have a major impact on growth patterns and level of density. Local governments typically are responsible for land use and growth management, and their decisions are vital in managing water resources. Through land use plans and ordinances, local governments can set development standards that determine where to build roads and infrastructure.

Land use planning and zoning practices have a significant impact on development patterns, which involve the layout of roadways. Land use is considered to be a vision of how a city or county wants to

FIGURE 4-7 Village in the Forest, a proposed five-thousand-acre development north of Atlanta, included a lake village as part of a mixed land use pattern. Image courtesy of EDAW.

grow; zoning defines the legal rights and uses for a piece of property. Smart land management is the key to sustainable growth, allowing both economic development and land conservation to exist side by side. In some states land use and zoning are required to be consistent. That means if land use changes, zoning changes with it.

Zoning determines where particular land uses are located, requirements for parking, sizes of roadways, permitted impervious land coverage, and the requirements for other physical elements. Zoning is also directly related to transportation patterns, with higher density zones being closer to interstates and arterials. Performance-based zoning, which is also referred to as bonus or incentive zoning, allows developers to increase density in some areas so as to preserve sensitive land in other areas. Overlay zones provide a greater level of restrictions in specific areas and may be used to influence a specific type of development or protect sensitive environmental resources.

Designing green roadways means looking beyond the road to a transportation approach that includes public transit as an integral piece of the mobility network. Communities struggle to accommodate growth in ways that meet environmental goals and reduce our dependency on automobiles. Many communities have turned to a concept called smart growth to make better decisions about planning where we live (Sipes, 2006).

Ten main principles of smart growth are:

- Create a range of housing opportunities and choices.
- Create walkable neighborhoods.
- Encourage community and stakeholder collaboration.

FIGURE 4-8 Small neighborhood parks are a part of the pedestrian-oriented development of Village at the Summit, a private community located near Snoqualmie Pass, Washington. Image courtesy of www.pedbikeimages.org / Dan Burden.

- Foster distinctive, attractive places with a strong sense of place.
- Make development decisions predictable, fair, and cost-effective.
- Mix land use.
- Preserve open space, farmland, natural beauty, and environmental areas.
- Provide a variety of transportation choices.
- Strengthen and direct development toward existing communities.
- Take advantage of compact building design.

Green roadways and smart growth go hand in hand. To create more livable neighborhoods, we need to make sure that our communities are designed to address the needs of all residents and improve their quality of life. That means taking into account the needs of a diverse population: young and old, wealthy and poor, healthy and disabled. It also means laying out a street pattern that helps create a sense of community.

Smart growth means better planning and more land preserved. Smart growth means condos and town houses mixed with single-family homes. Smart growth homes are closer together and linked by walking paths that lead to stores, movie theaters, offices, and mass transit. In many ways, smart growth advocates a return to the time before World War II, when we designed communities around people, not around automobiles. This means fewer streets, more grids,

FIGURE 4-9 Atlantic Station is a 138-acre, mixed-use development near downtown Atlanta, Georgia. It includes residences, office towers, and retail establishments within walking distance of Georgia Tech University. Image courtesy of J. Sipes.

and fewer dead-end cul-de-sacs, and sidewalks that encourage walkability.

The Congress for New Urbanism advocates for designing communities that emphasize many smart growth principles. These communities also promote the concept of green roads. New Urbanists want regions of thriving neighborhoods, connected by efficient, effective transit. They want neighborhoods that feel alive, where people from all walks of life can cross each other's paths and meet their needs. Street grids define the framework to allow this to happen because they slow down traffic, minimize congestion, and encourage walkability.

How we lay out our cities and communities has a significant impact on transportation systems. What does smart growth mean for transportation? According to the FHWA (www.fhwa.dot.gov/planning/sgindex.htm), it can mean:

- Establishing state and local land use strategies to increase population and housing densities and make transit more viable
- Managing and operating existing highway, transit, and other transportation modes to maintain or improve performance for each mode without adversely affecting neighborhoods or urban centers
- Knitting transportation improvement projects and public/private investments so that they merge as seamlessly as possible into the community

FIGURE 4-10 Atlanta's 15th Street, which is located just north of downtown off of Peachtree Street, is an example of an urban street that emphasizes pedestrian-scale uses. Image courtesy of EDAW.

- Supporting the provision of mixed use development so that transit, bicycle, and pedestrian facilities, and ferry boats are viable options to driving
- Accommodating the flow of freight throughout the country so that the economy can continue to grow

The goals and objectives of the smart growth movement are similar to those of CSS. More than anything else, they are both about making good decisions that consider what is best for people as well as the environment. According to the National Complete Streets Coalition (www.completestreets.org), in Washington, D.C., the most vibrant and desirable neighborhoods are within walking distance of the city's Metro stops. In Chicago, homes within a half-mile of a suburban rail station sell for approximately $36,000 more than houses farther away.

Reconnecting America, which advocates for mass transit, identifies $248 billion in transit developments that have already been proposed. Denver and Salt Lake City are each extending light-rail and bus lines into the outlying suburbs at a cost of approximately $5 billion per city. Tucson, Arizona, is developing a 3.9-mile streetcar system as an alternative mode of transportation for the downtown area. The streetcar project is part of the county's $2.1 billion Regional Transportation Plan that is funded by the Regional Transportation Authority through federal and other regional funds. Cost estimates for the project are $180 million.

The pattern of streets influences trip length. In a neighborhood with good connectivity, often achieved through a street grid, there are more intersections and thus more route choices. The traditional urban grid has short blocks, straight streets, and a crosshatched pattern. Traditional street grids disperse traffic rather than concentrating it at a handful of intersections. They offer more direct routes and hence generate fewer vehicular miles of travel than cul-de-sac networks. In contrast, the typical conventional suburban street network has large blocks, curving streets, and a branching pattern. Conventional suburbs tend to have fewer street connections, favoring isolated cul-de-sacs and looping roads that limit choices and cause congestion.

Street grids encourage walking and biking by offering direct routes and travel alternatives parallel to high-volume streets. According to Walkable Streets executive director Dom Nozzi, the shift from car trips to transit and walking occurs once certain job and housing densities are achieved.

The basic idea behind complete streets is that the roadway and adjacent areas are designed and operated so they work for all users, including pedestrians, bicyclists, motorists, and transit riders of all ages and abilities. A complete street may include sidewalks, bike lanes, bus lanes, transit stops, crosswalks, landscaped medians, pedestrian signals, curb extensions, and other features.

Some general principles of complete streets include the following:

- Understanding and application of context
- Flexibility and creativity to shape transportation solutions
- Preservation and enhancement of community and natural environments
- Consideration of multiple modes of transportation
- Creation of streets that are safe, accessible, and livable
- Humanization of the street, transforming it into a destination
- Proactive stakeholder involvement and interdisciplinary team
- Development of a unique solution (Sousa and Rosales, 2010)

Tools that can be used in achieving a complete street include road diets, traffic calming, intersection design, designing for pedestrians and bicyclists, transit design features, lane restrictions, and green street options.

FIGURE 4-11 In Madison, Wisconsin, the scale of development encourages pedestrian use adjacent to local streets. Image courtesy of www.pedbikeimages.org / Dan Burden.

Road Diets

A road diet involves removing travel lanes from an existing roadway and using the additional space for on-street parking, landscaped areas, sidewalks, trails, and other amenities. A common road diet is to convert a four-lane undivided roadway into two travel lanes and a center turn lane. Where a turn lane is not needed, the center can be converted to a raised median with ornamental plantings, or a lowered median used as a bioswale.

Another approach is to decrease the width of a lane to reduce vehicle speeds and provide space for other uses. A common approach is to reduce the width from twelve feet to ten feet in residential areas and then reduce the posted speed limit. The gained space can be used for bike lanes, roadside rain gardens, or street tree planting zones.

Road diets have generated benefits for all modes of transportation, including transit, bicyclists, pedestrians, and motorists. These benefits include reduced vehicle speeds, improved mobility and access, reduced collisions and injuries, and improved livability and quality of life.

According to a study conducted by the Highway Safety Information System, a multistate safety database, road diets have minimal effects on vehicle capacity because left-turning vehicles are moved into a common two-way left-turn lane.

Transit-Oriented Development and Highway Interchange Transit-Oriented Development

Transit-oriented development involves the creation of compact, walkable communities centered around high quality train systems. The idea is that you can walk to public transportation that takes you to work, to shop, or to play. As transportation costs rise, this type of approach is becoming more desirable. The benefits of a transit-oriented development are a high quality of life for people who live in such developments, an increase in public transportation use, and a reduced demand on the rest of the transportation system.

Highway interchange transit-oriented developments (HITODs) are highway interchanges that include real estate development within the highway right-of-way. The objective of HITODs is to build retail/commercial centers at selected interchanges as part of a public/private partnership. Revenue generated from this type of development is typically shared between a transportation agency and a private developer.

The City of Irving, Texas, has a HITOD that includes urban commercial, retail, and community development at the crossroads of State Highway 114, State Highway 183, and Loop 12, and is just minutes from I-35. The location of the HITOD makes it easily accessible to most of the Dallas-Fort Worth Metroplex.

Highway Corridor Overlay District

A highway corridor overlay district is a set of zoning regulations for parcels adjacent to major roads. These regulations can be instrumental in helping to create green roadways.

Value Engineering

It is common practice for transportation agencies to perform value engineering (VE) studies prior to construction or bidding. VE, which seeks to reduce the cost of a transportation project, is a part of the planning process for all major transportation projects. VE has been applied to transportation projects for years. The Federal-Aid Act of 1970 required VE and cost-reduction analyses on all federal-aid projects, and subsequent legislation requires the use of VE by all federal agencies.

One concern is that VE does not adequately take into account design features that make a transportation project more attractive, sustainable, and environmentally friendly, and can potentially result in cost reductions as well. Several states have noted that an unintended result of the VE process is the removal of items that are important to stakeholders.

The best way to ensure that the VE process does not lose sight of cultural, natural, visual, and historic concerns is to have landscape architects, historians, and key stakeholders involved.

Transportation Challenges

Meeting America's transportation needs will require a multimodal approach that preserves what has been built to date, improves system performance, and provides for other means of travel, including public transit, biking, and trails.

Anything we can do to get people to park their automobiles is helpful. Simply increasing bicycling from 1 percent to 1.5 percent of all trips in the United States would save 462 million gallons of gas each year.

Problems with Current Mass Transit

One major problem in the United States is the emphasis on individually owned automobiles. If you watch rush hour in any major city, the vast number of cars on the road have only one occupant. This isn't a very efficient way to handle transportation, and it isn't sustainable. Congress began taking steps toward the creation of a multimodal transportation system with the Intermodal Surface Transportation Efficiency Act in 1991, which increased investment in transit and nonmotorized transportation. However, nearly half of all Americans lack any access to mass transit, and only 20 percent live near high-capacity outlets for rail or rapid bus (Doster and Sheppard, 2009). In addition, an October 2008 survey by the American Public Transportation Association found that 85 percent of public transit systems reported capacity problems and 35 percent were considering service cuts. A 2011 study by the same association found that ridership on public buses and trains increased 2 percent—from 7.63 billion rides to 7.76 billion—for that year.

As transportation costs increase there may be a greater emphasis on public transportation, and cities that have been talking about light-rail and other

Figure 4-12 Wide, paved trails that are part of Alabama's Eastern Shore Trail allow cyclists to ride in a parklike setting without having to worry about highway traffic. Photo by Colette Boehm, courtesy of National Scenic Byways Online (www.byways.org).

Figure 4-13 Portland's Metropolitan Area Express (MAX) light-rail system helps reduce traffic demand by giving local residents an attractive, cost-effective alternative transportation approach. Image acquired from depositphotos.com.

FIGURE 4-14 The Portland Streetcar runs about every thirteen minutes during the day and costs about five dollars for a daily pass. Image courtesy of J. Sipes.

mass transit options may finally decide that it is better to spend their money via this approach than to build more roads. A problem with this idea, though, is that light-rail and other similar projects are expensive, and the federal funds used for their development are being cut along with all other transportation funds. Some cities have actually put mass transit projects on hold because of concerns for adequate funding.

One major problem is that mass transit has not been a priority in this country. According to a 2011 World Economic Forum study, over the past decade America's infrastructure has gotten much worse in comparison with other countries. For example, while the United States is investing $8 billion in high-speed rail, China is investing over a trillion dollars to create a nationwide high-speed rail network. In addition, China has opened a new subway system in each of the past six years. And France spends twenty times as much per capita on rail as the United States does. Despite the tough economic times of 2011, China is still on schedule to construct more than seven thousand miles of high-speed rail track in the next two to three years.

Since 1982, 80 percent of the U.S. Federal Highway Trust Fund has funded roads, and the other 20 percent has funded mass transit. The limited funds that have been earmarked for mass transit are used more for maintenance and upkeep instead of expanding opportunities.

Rising Interest

From 1995 through 2008, public transportation ridership increased by 38 percent—a growth rate higher than the 14 percent increase in U.S. population and higher than the 21 percent growth in the use of the nation's highways over the same period. In 2008, the American Public Transportation Association reported that public transit accounted for 10.7 billion passenger trips, which was the highest total in the past fifty-two years. One reason is that traffic congestion in the United States is considerably worse than it is in western Europe. Another is the high cost of owning and operating an automobile. The American Public Transportation Association estimates that an average household could save $9,499 a year by taking transit instead of driving. According to AASHTO, using transit results in savings of more than 3.9 million gallons of gasoline every day.

Thirty-five million times each weekday in the United States people board public transportation. Public transportation is a $48.4 billion industry that employs more than 380,000 people. According to the American Public Transportation Association, more than 7,700 organizations provide public transportation in the United States (www.apta.com/mediacenter/ptbenefits/Pages/FactSheet.aspx).

The Federal Highway Administration reports thirteen consecutive months of driving decline, with 112 billion less vehicle-miles traveled reported in early 2011 than in the previous thirteen-month span (Doster and Sheppard, 2009). This probably has as much to do with the high cost of gas as it does with frustration with congestion.

Alternative transportation is a critical part of the push for green roadways because the more we get people to park their cars, the fewer roads we have to build, and the more focused our efforts can be to create environmentally friendly roads.

Chapter 5

Public Involvement Process

Public involvement is a key component of virtually all successful transportation projects. Roads have such an impact upon our communities and landscapes, so it makes sense that people have a say in what happens with transportation projects. Public participation builds trust and support among planners, stakeholders, and local citizens. A green roadway project created with people rather than for them stands a better chance of benefiting everyone involved.

Creating green roadways involves initiating each transportation project in an atmosphere of collaboration and partnership. This approach helps ensure that all parties participate in defining a collective vision for a particular project.

The public participation process provides stakeholders with a forum for sharing knowledge of their communities, identifying opportunities and constraints, and providing input on priorities. As stated in the Federal Highway Administration's (FHWA) document *Flexibility in Highway Design* (1997), "Having a process that is open, includes public involvement, and fosters creative thinking is an essential part of achieving good design."

In the past, many transportation agencies did not always seek public input on projects. The traditional approach to transportation projects was for the engineers to make independent decisions using a standard design template, and then share the final decisions with the public. This approach was often referred to as "decide, announce, defend."

Fortunately, this approach no longer is acceptable. Stakeholders want to be involved with decisions made about any transportation project that has an impact upon where they live, work, or play. Federal legislation established basic principles for public involvement in transportation projects. In 1991, the Intermodal Surface Transportation Efficiency Act promoted greater opportunity for public involvement in the transportation planning process. The Transportation Equity Act for the 21st Century (1998) and the Safe, Accountable, Flexible, Efficient Transportation Equity Act: A Legacy for Users (2005) continued this approach. Public hearings are required by the FHWA for all highway projects that receive federal aid.

Public Involvement Begins at the Beginning

Public involvement in green roadway projects requires more than just setting up a website for the project or conducting a meeting or two. The objective is to implement a process that ensures stakeholders get an opportunity to be involved. The best way to develop a successful transportation project is to provide stakeholders an opportunity to get involved as early

FIGURE 5-1 In addition to the numerous meetings held with local leaders and community organizations, the local public is given an opportunity to participate with transportation projects. Image courtesy of J. Sipes.

as possible during the planning process. As a general rule, you don't want surprises at a public meeting, because they are seldom good surprises. Any community that suspects true dialog is not taking place is unlikely to support or contribute to a plan, however beneficial.

Public Involvement Plan

Developing a public involvement plan, which is typically the responsibility of the primary consultant for a transportation project, generally involves five steps:

- **Identifying stakeholders**–There are key stakeholders involved with every transportation project. These include adjacent landowners, people who participate in activities in the area, decision makers, and others who have a vested interest in the success of a project. Knowledge and understanding of a local community is critical in identifying stakeholders.
- **Interviewing stakeholders**–The next step is to conduct one-on-one interviews with stakeholders. Telephone or in-person formats can be used. Stakeholder interviews should cover a set of community issues, values, and constraints concerning the project. One approach is to add key stakeholders to a steering committee; this allows them to stay involved throughout all phases of a project.
- **Agency involvement**–An important part of any outreach effort for a transportation project involves discussions with agencies that need to be involved in the decision-making process. Because of their size and complexity, transportation projects typically involve numerous jurisdictions and agencies that are tasked with protecting cultural and natural resources. Early involvement can help identify issues the project is likely to face and can help determine what type of mitigation is likely to be needed to meet project objectives.
- **Selecting public involvement techniques**–There are a number of ways to get the public involved, including public meetings, websites, surveys, and newsletters. For projects involving diverse populations, materials need to be prepared in alternate languages to better serve non-English-speaking population segments.
- **Planning for implementation**–Once a project transitions into the implementation phase, public involvement changes, but is just as important as it was during the design and planning phase.

FIGURE 5-2 Scenic byways typically encourage people to stop, get out of their cars, and enjoy the scenery. This image shows birdwatchers taking advantage of the natural habitat that can be seen from a scenic overlook. Image courtesy of Lake Country Scenic Byway Association.

FIGURE 5-3 Bikers take advantage of the Ohio and Erie Canal Towpath Trail, one of Ohio's longest and most scenic bikeways. Image courtesy of National Park Service.

Box 5.1 Tools for Public Involvement

There is no shortage of publications and tools oriented toward giving the public a voice in public projects. Some of these include:

- The FHWA's **Innovations in Public Involvement for Transportation** consists of a set of nine leaflets that contain practical techniques of public involvement.
- The Transportation Research Board's Committee on Public Involvement has a website (https://sites.google.com/site/trbcommitteeada60/) that provides the transportation community with vital information on research, conferences, publications, training, tools, and techniques related to public involvement in transportation.
- The Transportation Research Board's Transit Cooperative Research Program
- **Synthesis 89: Public Participation Strategies for Transit** documents offer public participation strategies to inform and engage the public for transit-related activities.
- The FHWA and Federal Transit Administration created **A Guide to Transportation Decision Making** that is intended to help stakeholders understand how transportation decisions are made at the local, state, and national levels, and the opportunities available during the process to contribute ideas and concerns.
- The Virginia Department of Transportation has published **Public Involvement: Your Guide to Participating in the Transportation Planning and Programming Processes** to explain the planning and programming processes, which are the beginning stages of all transportation projects. The brochure also explains the various activities in which citizens can participate and influence plans years before project-level work begins.
- Oregon DOT's Transportation System Planning program involves early coordination with review agencies and local governments as well as creation of a public involvement program.
- The FHWA report **How to Identify and Engage Low-Literacy and Limited-English-Proficiency Populations in Transportation Decisionmaking** provides recommendations for working with low-literacy and limited-English-proficiency populations.
- In 1999, the Minnesota Department of Transportation finalized a public involvement process called **Hear Every Voice**. The process defined guidelines for how to develop public involvement plans.
- Washington State DOT's **Building Projects that Build Communities** handbook focuses on new and better ways of building highways based on a growing interest in the improvement of highways and their integration into the communities they serve.

Social Media and Public Involvement

Social media refers to web-based and mobile technologies that can be used to provide almost immediate input and feedback. What makes social media so powerful is that information can be made available virtually in real time, and is accessible to anyone with a computer, tablet, or smartphone. Social media can include Internet forums, blogs, wikis, podcasts, and other media approaches.

Platforms such as Facebook and Twitter are effective ways to keep people informed about a given project. It is becoming commonplace for transportation projects to have their own Facebook page. One benefit of a Facebook page is that you can see how many people visit your page, and keep track of how many people "like" the things you post online. It also provides an opportunity to learn more of the concerns and ideas of stakeholders for a given project.

Twitter is a social networking and microblogging service that is used to inform others of your specific actions. By setting up a Twitter account for a specific project, short bursts of information, called "tweets," about upcoming events, key decisions, and other information can quickly be sent to anyone who signs up to follow you.

Social networking is expected to continue to evolve as mobile, interactive technology continues to improve. Neighborhood America and Public Comment are both web-based programs intended to help improve communication with stakeholders, clients, consultants, and other team members.

Web-based tools for public participation are inexpensive, easy to access, and easy to use. In addition, web-based tools can offer greater access to groups that have difficulty reaching public participation meetings, such as the physically handicapped or the elderly, and can also allow individuals living in rural areas to participate in the decision-making process without traveling long distances to do so.

Participants at public meetings do not always accurately reflect community views, and the loudest participants frequently dominate a meeting. Individuals who are hesitant to express their opinion at a public meeting may find it more comfortable to offer feedback in an online forum.

Visualization

To strengthen public participation in the planning and project delivery process and specifically to aid the public in understanding proposed plans, visualization techniques can be used to help the public "see" what is being considered for a specific project.

The basic idea behind a visualization is to accurately convey what a transportation project would look like once it is implemented. Visualizations can be generated using computer graphics tools, hand-drawn graphics including sketches or paintings, or three-dimensional (3-D) cardboard models.

There is an increasing interest in using 3-D modeling tools to develop design and planning concepts for transportation projects. 3-D models can be used for a variety of different simulations and analyses, including environmental impact assessments, shadow studies, flood and other hazard modeling, visual impact simulations, and massing studies.

Because 3-D models can be so accurate, viewers can relate to them, and establishing this context helps when discussing proposed changes. All 3-D models can be enhanced by adding colors, textures, transparencies, or other characteristics to create more photo-realistic images.

Animation creates the illusion of motion by displaying a series of still images in rapid succession. Animation can be very time-consuming and expensive because of the effort it takes to produce high-quality imagery, but it can be very effective for transportation projects because the view from the road is best shown as if from a moving car.

Figure 5-4 Public meetings provide stakeholders an opportunity to share their thoughts and help influence decisions for green roadway projects. Image courtesy of J. Sipes.

FIGURE 5-5 With the aid of visualization techniques, a presenter can create images that provide the public with a better idea of how a given project could look once implemented. Image courtesy of AECOM.

Partnership Opportunities for Green Roadways

The way we have designed and constructed transportation projects in this country no longer works, and many state departments of transportation (DOTs) are looking for new, innovative approaches to get the job done. One approach is to develop partnerships with other state or local agencies, nonprofit organizations, academic institutions, and private developers to meet transportation objectives.

Partnerships are an effective way to get key stakeholders involved in a transportation project. It is also a great way to stretch the budget for a project. For example, a conservation group may be willing to purchase additional land adjacent to a roadway right-of-way to protect natural resources, or they may be willing to be responsible for landscaping with native materials or for environmental restoration projects associated with a given roadway project.

Any organization or agency interested in a specific transportation project is a potential partner. This partnership may be public, private, or a combination of the two. Examples of innovative partnerships include the following:

- **Trees Forever**, a nonprofit organization, works with the Iowa DOT to administer the state's Living Roadways program. Trees Forever is actively involved with planting and maintaining plant material along the state's highways.
- The **Virginia Green Highway Initiative** is a partnership among the Virginia Tech

Transportation Institute; Virginia Tech's Institute for Critical Technology and Applied Science, College of Engineering, and Office of the Vice President for Research; and the Virginia DOT. The goals of the initiative are to increase energy efficiency and reduce carbon emissions, emphasize green energy, and minimize negative impacts on Virginia's ecosystems.

- The **Green Highways Partnership** was initiated in 2002 through the combined efforts of the U.S. Environmental Protection Agency and FHWA. The Green Highways Partnership is a voluntary public-private collaborative that advances environmental stewardship in transportation planning, design, construction, operations, and maintenance while balancing economic and social objectives.
- The Washington State DOT's **Transportation Innovative Partnership Program** is a formal process for the state to evaluate transportation public/private partnership proposals.
- The Oregon DOT partnered with Portland General Electric, the state's largest utility, to complete the nation's first solar photovoltaic project in a highway right-of-way as part of the department's Solar Highway project.
- Many municipalities have some type of Adopt-a-Highway and/or Partnership Planting Program that allows private citizens and organizations to get involved with maintaining a segment of roadway.

Case Study–Public Participation

Vancouver Land Bridge, Vancouver, Washington

The Vancouver Land Bridge is a great example of how a green transportation project can not only improve pedestrian access and connectivity, but also highlight the history and culture of a place. The Land Bridge is a pedestrian bridge that spans State Route 14 just east of Vancouver, Washington, connecting the historic site of old Fort Vancouver with the Columbia River waterfront. But the land bridge is much more than that. It also commemorates the confluence of rivers and indigenous people encountered by the Lewis and Clark expedition. The land bridge is one of seven sites that make up the Confluence Project, which traces 450 miles of Lewis and Clark's route west. The Confluence Project involved participation of stakeholders throughout the Columbia River basin.

The land bridge was designed by Jones & Jones, a Seattle architecture and landscape architecture firm. The bridge and the approach ramps follow a long, sweeping curve that makes it seem as if the land is just flowing from one side of the road to the other. Three overlooks along the bridge provide views of surrounding features, including Kanaka Village, Fort Vancouver, the Columbia River Waterfront, Mount Hood, and the Cascade Mountains. Seating opportunities along the bridge provide places to rest and reflect, and interpretive signage tells the story of Lewis and Clark, the Klickitat Trail, and the relationship of American Indian cultures to the land. More than 3,100 linear feet of trails create the meandering, serpentine character of the project.

The bridge was designed to capture all stormwater runoff and direct it to a rain garden on one side and a human-made creek on the other. The use of indigenous plants for the landscaping reduced the need for irrigation, and a shallow groundwater well fed by the Columbia River aquifer was used for irrigation while plants were being established.

Construction of the land bridge was complicated because the design of the project had to meet expectations of various American Indian tribes while also accommodating the severe physical constraints of the site. The design team worked closely with the City of Vancouver, Washington State DOT, the Federal Aviation Administration, the BNSF Railway Company, and the National Park Service to coordinate all of the critical issues that needed to be addressed.

Construction of the Vancouver Land Bridge project, which totaled $12.25 million, was made possible through a partnership of Confluence Project members, the National Park Service, the City of Vancouver, and the Washington State Department

Figure 5-6 The Vancouver Land Bridge provides a safe pedestrian route along the Renaissance Trail that runs along the Columbia River. Image courtesy of Jones & Jones.

Figure 5-7 As bikers and pedestrians cross the Vancouver Land Bridge, they experience what is intended to be a natural extension of the landscape. Image courtesy of Jones & Jones.

Figure 5-8 At the grand opening of the land bridge in August 2008, an inaugural "first walk" was conducted over the bridge. Image courtesy of Jones & Jones.

Figure 5-9 Visitors attend the grand opening of the Vancouver Land Bridge. Image courtesy of Jones & Jones.

FIGURE 5-10 Puppets from the Lion King were part of the grand opening festivities of the land bridge. Image courtesy of Jones & Jones.

of Commerce, with additional funding coming from federal, state, and private sources.

THE CONFLUENCE PROJECT

The Confluence Project was initiated in 2000 through a collaboration of Pacific Northwest American Indian tribes and civic groups from Washington and Oregon. From the very beginning, founders of the project wanted an inclusive process that would encourage participation. More than forty organizations collaborated in the process, and dozens of government agencies (federal, two states, and countless counties and cities), nonprofit organizations, and American Indian tribes have been involved.

The group invited renowned artist Maya Lin to develop an art installation for each of the seven sites that would interpret the area's ecology and history, and reference a passage from the Lewis and Clark journals. Lin worked with civic groups from Washington and Oregon, as well as with other artists, architects, and landscape designers.

Jane Jacobsen, executive director of the Confluence Project, said, "The Confluence Project is about bringing people together for a better future, and it's made possible by a solid team effort."

A number of outreach efforts continue to promote the Confluence Project. An e-newsletter is available, there is a Facebook page for the project, and a Flickr page allows people to post photographs they have taken of their visits to Confluence sites. An online Journey Book offers a way to explore each of the sites that make up the Confluence Project. Confluence Project in the Schools is a school-based arts education program that links students and teachers with professional artists, tribes, and community partners. Gifts from Our Ancestors is a two-year program led by the Confluence Project, local artists, and educators. The goal of the project is to help ensure the continued educational, cultural, and ecological stewardship of the Columbia River and its tributaries.

The total cost of the Confluence Project is expected to be around $27 million.

Chapter 6

Green Roadways in Urban Areas

Now more than ever, stakeholders in urban areas want to be involved in shaping transportation infrastructure to be compatible with community objectives. For green roadways in urban areas, streets should enhance mobility choices and complement and reinforce community character, livability, and sustainability.

Deemphasizing Roads

One of the fundamental problems in urban areas is that there is already too much paving, which causes significant environmental problems, including increased stormwater runoff, flooding, water and air pollution, loss of open space and wildlife habitat, and increased temperature. Urban areas are hotter than they otherwise would be because paved surfaces absorb and radiate solar energy. On average, U.S. cities are two to eight degrees Fahrenheit warmer than undeveloped areas.

Excessive paving in urban areas also has a negative impact on quality of life. Roads have created barriers that divide neighborhoods and make it difficult to walk or cycle. According to a study by the U.S. General Accounting Office called "Traffic Congestion—Road Pricing Can Help Reduce Congestion, but Equity Concerns May Grow" (2012), the average annual commuting delay has more than tripled in the last decade, and commutes are likely to get even longer as cities expand. Solving this transportation problem in urban areas is not about building more roads.

Communities struggle to accommodate growth in ways that meet environmental goals and reduce our dependency on automobiles. Many people would like to live in walkable, mixed-use neighborhoods that are closer to their jobs, so they can spend more time with family and less time driving. (See Land Use Planning, Smart Growth, Complete Streets, and Transportation Infrastructure in chapter 4.)

A single commuter switching his or her commute to public transportation can reduce a household's carbon emissions by 10 percent. The reduction can increase to 30 percent if he or she eliminates the household's second car. When compared to other household actions that limit carbon dioxide emissions, taking public transportation can be ten times greater in reducing this harmful greenhouse gas.

Retrofitting Existing Streets

With so many of our older roads nearing the end of their life expectancy, federal, states and local governments must decide if it is smarter and less expensive to renovate highways or to build new routes. More often than not, in urban areas the decision is made

FIGURE 6-1 Shoppers in downtown Greeley, Colorado, feel a sense of community as they enjoy the many opportunities of the city. Image courtesy of AECOM.

FIGURE 6-2 The Plaza Arts Center in Atlanta, Georgia, serves as an inviting and comfortable community gathering place for many activities. Image courtesy of AECOM.

Figure 6-3 This wide street in Barcelona, Spain, is paved with patterned paving stones and lined with ornate wrought iron street lamps. Image courtesy of www.pedbikeimages.org / Charlie Zegeer.

Figure 6-4 This bicycle intersection in Newport Beach, California, allows commuters to safely travel to their destination, avoiding traffic concerns. Image courtesy of www.pedbikeimages.org / Dan Burden.

Figure 6-5 In Seattle, Washington, metal railings help create a sense of separation from the road, and are also used to hold up decorative planters. Image courtesy of www.pedbikeimages.org / Dan Burden.

to retrofit existing streets because there typically isn't space to build new ones. The Complete Streets program is a good example of the emphasis on retrofitting existing streets (see chapter 4).

Retrofit-focused roadway projects usually involve repairing existing streets or adding lanes in an attempt to reduce congestion. In some cases, counter to intuition, they involve removing existing streets or implementing a road diet to reduce lanes and place more of emphasis on public transit and pedestrian circulation.

Retrofitting existing roads provides an opportunity to incorporate green roadway standards and create more environmentally friendly landscapes and more walkable streets.

Figure 6-6 After it was retrofitted as part of a streetscape enhancement effort, Peachtree Street in Atlanta, Georgia, allows more walking and cycling, transforming Atlanta into a more pedestrian-friendly city. Image courtesy of AECOM.

Figure 6-7 The Northeast Siskiyou Green Street curb extensions include a series of miniature dams. Water cascades from one "cell" to another. The system can manage 225,000 gallons of stormwater runoff per year. Image courtesy of Kevin Robert Perry, City of Portland.

Figure 6-8 The Woodlands Waterway in Texas is a 1.4-mile–long corridor that provides residents and tourists a fun way to get around the Woodlands Town Center. Image courtesy of AECOM.

In Lakewood, Colorado, for example, a retrofit of an outdated low-rise shopping mall resulted in a mixed-use development that emphasizes walkability based on an interconnected street grid. The acres of parking have been replaced by ground-level retail space topped by offices and residences.

Many cities have changed, or are looking to change, their existing roadways:

- In 1973, New York City demolished the decrepit West Side Highway. As a result the neighborhoods of Chelsea, Tribeca, and West Village have thrived.
- In San Francisco, the elevated Embarcadero Freeway was torn down when it was damaged in an earthquake in 1989.
- Boston's Big Dig relocated streets in the downtown area underground, freeing up the surface for parks and surface streets.
- Removal of roads in downtown Portland, Oregon, helped open up more than three hundred acres of land and served as the catalyst in redevelopment of the downtown waterfront.
- Oklahoma City is removing the Crosstown Expressway from the downtown area and installing a new ten-lane highway that is located in a trench. Tree-lined city boulevards run over the top of the new highway, and eighty city blocks are being redeveloped along the Oklahoma River.
- Cleveland wants to convert its West Shoreway, next to Lake Erie, from a fifty-mile-per-hour freeway into a tree-lined boulevard.
- The Congress for New Urbanism and the Center for Neighborhood Technology are promoting a Highways to Boulevards Initiative.
- Nashville wants to replace eight miles of interstate—parts of I-65, I-40, and I-24—with parks and neighborhood streets.
- Buffalo wants to get rid of its Skyway, an elevated highway that blocks access to Lake Erie.
- Akron, Ohio, conducted a study to investigate tearing down its 2.2-mile Innerbelt that leads downtown from I-76/I-77.

New Transportation Projects

There aren't as many new highway projects in urban areas because of the lack of available space, but when such a project is initiated, it typically affects public open space or requires the relocation of local businesses and residents. Unfortunately, when residential areas are affected, low income or minority neighborhoods often take the biggest hit.

For any new highway project under consideration, there needs to be a discussion about the feasibility of the project and how best to make sure it fits with surrounding land uses. Many cities that constructed bypasses probably would make a different decision if they could do it over again because of the negative impact the bypasses had on downtown areas. Many cities are opting to tear down or move below grade surface roads and elevated highways to free up space and enhance community character.

Some cities are concentrating on building more local streets instead of highways. Local streets handle approximately 50 percent of all vehicle miles traveled in the United States. The Congress for New Urbanism believes that the most efficient approach to transportation is a highly networked system of streets with at least 150 intersections per square mile because it has redundant routes, compact block sizes, sidewalks, and narrower streets.

Although most of the new transportation projects in urban areas still focus on building roads, that trend seems to be shifting. Many cities are looking at light-rail and other mass transit projects to address traffic problems. Because it isn't possible to build enough roads to meet transportation demands, transit-oriented projects may soon be more popular than road projects.

Urban Case Studies

Mexicantown Bagley Avenue Pedestrian Bridge and Plaza, Detroit, Michigan

The Mexicantown Bagley Avenue Pedestrian Bridge and Plaza is intended not only to improve connectivity, but also to reconnect Detroit's Mexicantown community. Mexicantown is one of southwest Detroit's oldest neighborhoods, but it was divided into two sections when I-75 was built in the late 1970s. The pedestrian bridge reconnects the two sections of Mexicantown. The bridge, which was completed in 2010 at a cost of $5 million, is part of the $230 million Michigan Department of Transportation's Gateway Project, which provides a direct connection from I-75 and I-96 to the Ambassador Bridge. The project is a good example of how green roadway projects can promote walkability and reconnect communities. The Gateway Project is expected to increase commerce by supporting more than six hundred existing businesses and attracting new businesses, supporting economic development, and enhancing tourism opportunities.

The 407-foot-long bridge spans I-75 and I-96, two of southeast Michigan's busiest freeways. The bridge is the first cable-stayed pedestrian bridge in Michigan. A single 150-foot tower provides support, and the bridge varies in width from ten feet on the western approach to thirty-one feet on the eastern approach. An urban plaza is located on each end of the bridge.

Noted Michigan artist Hubert Massey created two works of art to celebrate Mexicantown's heritage. The first work, entitled *Spiral of Life,* is a forty-foot-long and five-feet-high mosaic of handcrafted tile that traces the history and culture of the people who live in Mexicantown. The second work, called *The Spiral Kinship,* is a twelve-foot-tall vertical form made of textured aluminum and featuring a bronze-colored globe mounted on top. The art is part of a plaza that serves as the eastern entrance to the bridge.

Boston's Big Dig

When it comes to redevelopment projects, they don't get much bigger than Boston's Big Dig. The project started in the 1970s and was completed in December 2007, the month the latest recession began. The Big Dig was the most ambitious urban infrastructure

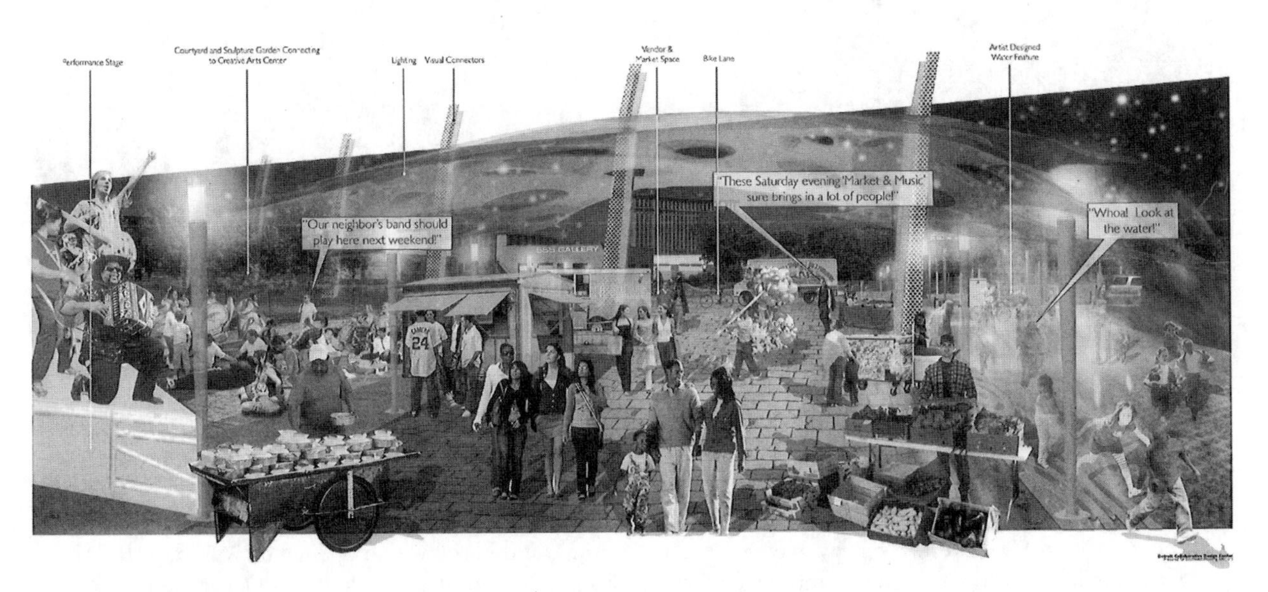

FIGURE 6-9 Mexicantown in southwest Detroit has a unique character that attracts visitors from all over the state. Image courtesy of Michigan Department of Transportation.

Figure 6-10 The Mexicantown Bagley Avenue Pedestrian Bridge connects to plazas at both ends. Image courtesy of Michigan Department of Transportation.

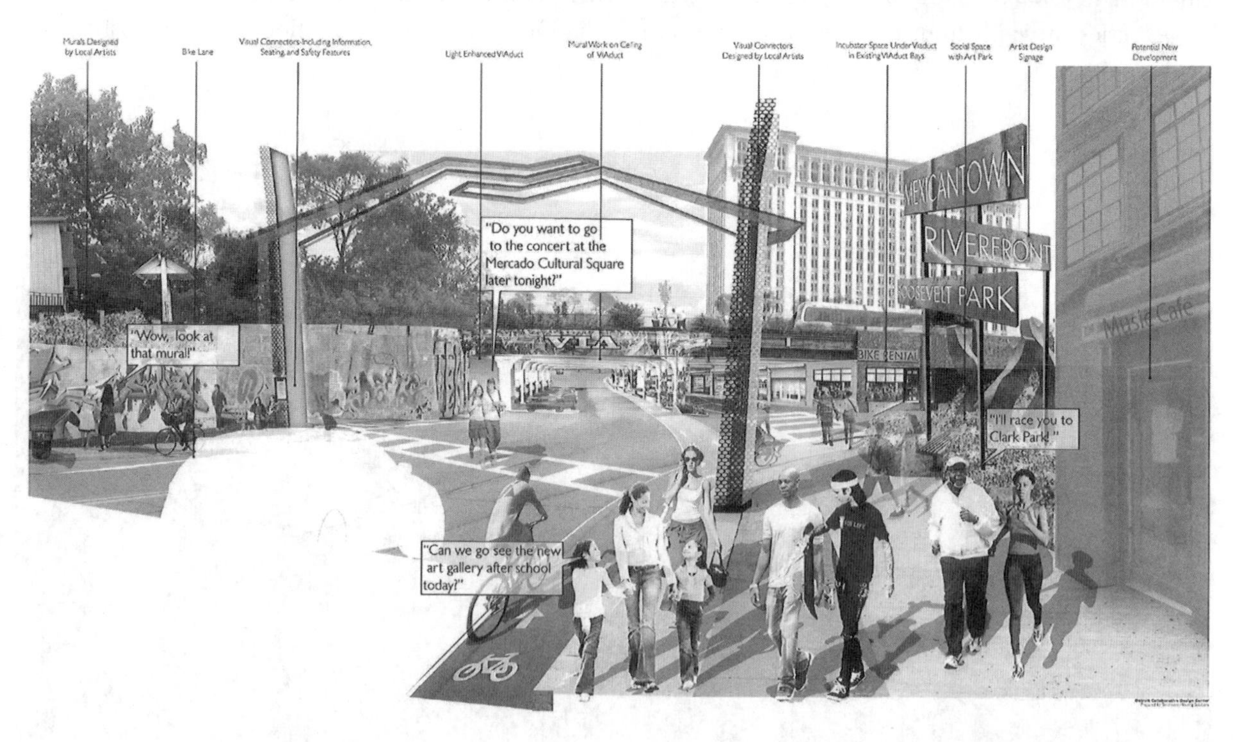

Figure 6-11 The Mexicantown Bagley Avenue Pedestrian Bridge and Plaza, which opened on May 5, 2010, is a vital link in the Mexicantown community Gateway Project. Image courtesy of Michigan Department of Transportation.

project in the United States in decades. There is much debate over the effectiveness of the project, but it has definitely removed traffic from downtown Boston streets, reduced congestion, improved community character, and increased available public space, which is leading to economic revitalization.

The total price for the Big Dig is estimated to be around $22 billion, making it one of the most expensive public works projects in the nation's history. For perspective, the Commonwealth of Massachusetts is paying more than $600 million a year for the project, and the loan will not be paid off until 2038. There are a lot of questions about how the project was funded. Federal funding was obtained when Massachusetts Democrat Tip O'Neill, the House's majority leader, convinced Congress to insert a "placeholder" for project funds by trading favors with other congressmen who wanted "pork" projects for their states, too. But even with this placeholder, financing was a disaster. Original cost estimates for the Big Dig were around $2.6 billion, considerably less than the final cost, and project leaders were less than forthright as costs skyrocketed.

The project involved removal of an elevated highway through downtown Boston, building the new highway in tunnels, and developing the surface as a public green space. The old artery was built along Boston's waterfront in the 1930s, and ever since then it had separated the city from the water. The artery quickly became obsolete, and despite numerous projects to correct the problem, traffic in downtown Boston became one traffic jam after another.

Construction on the Big Dig started in 1991. Over the course of the project, canyons and tunnels were literally carved under Boston for 3.5 miles, and 7.5 miles of highway, six interchanges, and two hundred bridges were constructed.

FIGURE 6-12 The Big Dig was a project in Boston that rerouted I-93 into a 3.5-mile tunnel. The project was created in response to Boston's heavily congested streets. Image courtesy of Wikipedia Commons.

WHAT WENT WRONG

It shouldn't come as a surprise that there was no shortage of problems with a project as large and complex as the Big Dig. Some refer to Boston as a cautionary tale about how difficult it is to transform an urban city and redefine the role the automobile plays in creating urban form.

Mismanagement led to a number of mistakes, and quality control was sorely lacking. The project was so large that several construction companies formed a joint venture to handle the construction.

In 2006, a ceiling panel from one of the tunnels fell, killing a woman, and there is concern that other panels and ceiling lights are also unstable.

One of the biggest complaints is that officials in charge of the Big Dig were not honest with the public

FIGURE 6-13 Boston's Big Dig reconstruction replaced part of the highway with a mile-long green space, including benches, fountains, and trees. Image courtesy of Wikipedia Commons.

about the true costs of the project. Apparently during the first two years of construction, officials knew the original estimates were up to $6 billion off, but state managers didn't want to make the difference known to the public. The state eventually had to admit the funding problems to federal investigators, and there was public outrage about what some called fraudulent accounting.

WHAT WORKED

The project not only significantly improved traffic patterns, it completely transformed downtown Boston. A study by the Massachusetts Turnpike Authority determined that travel times through Boston dropped from 19.5 minutes to 2.8 minutes.

The state made the decision back in 1991 that 75 percent of the land created as a result of the Big Dig would be left as public open space. The old highway was torn out and replaced with a mile-long linear greenway, called the Rose Fitzgerald Kennedy Greenway, that promotes pedestrian movement. New buildings are being developed along the greenway, and many dilapidated buildings are being restored. Commercial properties have increased in value, and investors and residents seem to be optimistic about the future. One of the biggest challenges will be to follow through with developing the greenway parks and continued urban revitalization efforts.

Portland's Green Streets Program

The Willamette River, which runs through the heart of downtown Portland, Oregon, was consistently being polluted from nearby tributaries. As a result of a Clean Water Act lawsuit in the early 2000s, the city had to find a way to clean up the problem and improve water quality. One option to fix the problem was to add numerous drainage pipes and enlarge existing pipes, which would cost an estimated $150 million. But instead of this traditional engineering approach, the City of Portland decided to take a more environmentally friendly approach by implementing a Green Streets program. The initial cost for implementing and maintaining such a program was projected to be about $11 million, which was considerably less than the engineering approach. Even with later expansions, the total cost of the Green Streets program came out about $60 million less than the estimate for adding new pipes.

Between 2002 and 2005, the U.S. Environmental Protection Agency granted the city $2.6 million for more than twenty-five innovative public and private projects throughout the city that demonstrate sustainable, low-impact stormwater solutions.

In 2007, Portland City Council approved a policy to promote the use of green roadways in design and

Figure 6-14 Portland's Green Streets project sign helps inform residents of the benefits of sustainable stormwater management. Image courtesy of Kevin Robert Perry, City of Portland.

construction. This would include a multitude of different green improvements to the roadways, such as the following:

- Improved bicycle and pedestrians paths
- Increased trash and pollution pickup
- Additional green areas near roadways
- Diversion of water from storm sewer systems to avoid backups
- Recharged groundwater

Since then, Portland has been on the forefront of promoting green streets. More than five hundred green street facilities are already in place, with plans to add another five hundred in the next few years. There are many different types of green improvements on these streets, including:

- Replacing normal grass strips with vegetation, which reduces the flow of water and allows the roots of the plants to soak up more water.
- Planting large vegetation or something equivalent in large asphalt areas that are likely to flood.
- Painting green bicycle boxes at intersections to minimize conflicts between drivers and bikers. This green box allows bicyclists to come completely in front of any vehicles queuing up.
- Planting swales, planters, or basins along the roadway. These features would absorb a considerable amount of stormwater, sometimes as much as 100 percent.

Cities from around the country are looking to Portland as an example of how to implement natural, long-lasting green street solutions. The Green Streets program has made transportation safer for pedestrians, bicyclists, and commuters.

Multifunction Roundabout, Normal, Illinois

Most people have experienced driving through a roundabout; they are becoming more and more

FIGURE 6-15 The bronze sculptures in downtown Portland are an important part of the city's character. Image courtesy of www.pedbikeimages.org / Andy Hamilton.

FIGURE 6-16 The roundabout in Normal, Illinois, not only serves as a welcoming public gathering place, but also contains an underground rainwater collection system. Image courtesy of City of Normal, Illinios.

common in America. The town of Normal, Illinois, has created a roundabout that serves as more than just a roadway. By blending form and function, the roundabout in Normal manages traffic and treats stormwater runoff, but it is also a place to hang out, walk, visit with friends, or just sit and rest or enjoy the scenery.

The Normal roundabout sits at the intersection of five converging roads. It is two lanes and functions exactly as any other roundabout does—it facilitates traffic in a circular pattern instead of using signals. Traditionally, roundabouts have been designed so that the center of the roundabout is somewhat utilitarian. There may be some landscaping in the center of the roundabout, but the area is usually considered to be leftover space that doesn't really have a purpose.

The Normal roundabout is designed with a very different philosophy. Its center is actually an urban park that includes walkways, a grass lawn, landscaping, and seating opportunities. The circular walkway surrounding the park invites people in. A small bridge crosses a wet-bottomed bog that includes native grasses, shrubs, and trees. A reflecting pool encircles the lawn. But what really makes this roundabout interesting is what goes on beneath it.

The roundabout is actually a water treatment facility that captures and filters water from the downtown Normal area. Stormwater runoff from surrounding roads drains into the roundabout and the water is directed through a series of filters, and then into an existing 76,000-gallon storage tank or one of the cisterns underneath the roundabout. The water is also treated by an

FIGURE 6-17 The rainwater collection system in the roundabout collects runoff from two streets. The water is captured, stored, and recycled, making stormwater management a visible public amenity. Image courtesy of City of Normal, Illinois.

ultraviolet light that kills microorganisms without the use of chemicals to make the water clean and safe for animals and people when the water is used for irrigation purposes. This water is then pumped through the bog, which further filters out any remaining debris and toxins. The water is then gravity-fed into the four reflecting pools throughout the roundabout and the fountain. After the water is filtered through this system, it is held in a cistern where it can be used to irrigate adjacent areas.

Delaware Avenue Expansion, Philadelphia, Pennsylvania

The City of Philadelphia is proposing to extend Delaware Avenue by about one mile as part of the master plan for the development of the Delaware River waterfront. The project would extend Delaware Avenue from Lewis Street over Frankford Creek and under the Betsy Ross Bridge to Buckius Street. This will be the first new road constructed along Philadelphia's Delaware riverfront in decades. It is estimated to cost $18 million.

This new road would provide access to new development and recreational opportunities in the area. For example, the site of the former Dodge Steel plant is being transformed from a vacant brownfield site to a major public/private development that is expected to include new condominiums, a public park, and a multiuse riverside trail.

The project originally started in early 2004, but it was much later that sustainability was emphasized and the redesign of Larder's Point Park was included in the discussions. The design for the road also incorporated stormwater management regulations that had recently been enacted by the City of Philadelphia. Vegetated swales are used to manage stormwater runoff, which allows the first inch of runoff to infiltrate into the ground. The road surface is super-elevated so that stormwater is pitched in one direction to allow all the runoff to flow into the vegetated swales alongside the road. This approach greatly reduces the amount of stormwater treatment facilities needed for the project, resulting in significant cost savings.

There has been a dispute between the City of Philadelphia and the Delaware River Port Authority (DRPA) over which entity would be responsible for paying $2 million to make improvements along Hedley Street. The city needs an easement from DRPA because the avenue would go under the Betsy Ross Bridge, which belongs to DRPA. In return for the easement, DRPA wants the city to pay for repairs to Hedley Street.

T-REX Light-Rail Pedestrian Bridge, Denver, Colorado

The Transportation Expansion (T-REX) Multi-Modal Transportation Project involved the development of nineteen miles of new light-rail transit lines and improvements to seventeen miles of two interstate highways in the metro Denver area. One of the most innovative parts of the project is that it was constructed using a design-build process that allowed one contractor team to design and build the entire project.

Figure 6-18 The T-REX project in Denver, Colorado, is unique in that it combines a light-rail, a highway, and a pedestrian and bike trail, while also addressing traffic issues. Image courtesy of AECOM.

The project was initiated in 1992 to look at congestion in the Denver region, and later became known as the Southeast Corridor Major Investment Study. The study recommended a combination of new light-rail, highway improvements, pedestrian and bicycle facility improvements, and enhanced travel demand management measures. A preferred alternative, which included a new light-rail system and various highway improvements, was selected in early 1999 during the draft environmental impact statement process.

The Colorado Department of Transportation and the Regional Transportation District worked jointly to obtain federal and local funding for the project. The benefit of this approach is that it expedited the construction process, was more cost-effective and flexible, and led to creative, innovative solutions to some difficult construction problems. Using the design-build method, the contractor was able to complete the project twenty-two months ahead of the original schedule. The light-rail component cost $879 million, and the highway component cost $795 million.

The light-rail trains operate on double-track lines that are continuously welded and rest on concrete ties for a smooth and quiet ride. Several stations are being integrated with ongoing transit-oriented development, which are compact, mixed-use developments located around the transit stops. A self-service fare collection method is located at each station along the rail system.

A number of bridges along the I-25 corridor were replaced during T-REX construction. In partnership with the local municipalities along the corridor, T-REX erected six pedestrian bridges at light-rail stations.

Stormwater runoff was a major problem in the southeastern part of the corridor. Facilities designed to remove pollutants and enhance stormwater quality were built at interchanges. Other stormwater improvements include a new box culvert storm sewer trunk main, a thirteen-foot tunnel that carries stormwater, and additional storm sewers along streets.

Intelligent transportation systems were used for the parking management system, light-rail station traveler information system, and transit signal priority. Seven dynamic message signs within the project area and another six within the region keep motorists informed about traffic. As part of the project a new supervisory control and data acquisition system was installed that allows continuous monitoring and control of the system from a central location.

One reason the project was successful is the extensive process of public and agency involvement. Public meetings were held to get an idea of what stakeholders wanted, and many of those ideas were implemented. Public art was also incorporated into the transit stations as a way to help create a stronger connection to the neighboring communities. Artists developed concepts for railings, seating, and shelters at each station.

The T-REX project has increased transit options, enhanced safety for motorists, replaced aging infrastructure, improved mobility, and supported residential and commercial areas along the corridor.

Summary

Urban areas are significantly affected by the development of roads, parking lots, and other transportation-related facilities. The problem in most urban areas is that the landscape is so automobile-oriented that community character and quality of life have deteriorated. The push toward green roadway projects is intended to find more of a balance that meets transportation needs while also supporting community character, livability, and economic and environmental sustainability.

Chapter 7

Green Roadways in Rural and Suburban Areas

Rural and suburban areas both depend heavily upon the automobile, and trying to find a balance of meeting transportation demands while protecting natural resources and quality of life has proven to be difficult for many communities. As explained in chapter 6, urban areas already have a strong network of roads but limited space for expansion or modification. In less densely inhabited communities, there is more open space for roads, and yet many miles of road may need to be built to serve only a small number of people. Accommodating the loss of efficiency of scale while increasing transportation access and balancing natural resource protection are the major issues of suburban and rural transportation.

Rural Transportation

Rural areas near large urban areas often struggle to keep up with growth and traffic demands. Many are struggling to limit growth and do not encourage transportation projects because those often just encourage growth. It is difficult for these rural areas to maintain their visual and cultural character, and often they are incorporated into the neighboring city and quickly become suburbs. Other rural areas establish land use and zoning plans that actively control growth, so development areas grow slowly, existing natural resources are protected, and green infrastructure is added in a more organized manner.

Other rural areas have experienced declines in population as people have migrated to urban centers to seek employment. As a result, the rural population is becoming disproportionately elderly as younger people are moving to urban areas. Rural areas generally support fewer jobs than metropolitan areas, and thus, those workers who remain in their rural homes must drive increasingly longer distances to get to work.

According to the Federal Highway Administration (FHWA), rural America makes up more than 80 percent of the nation's land and includes more than 3.1 million miles of rural roads. Rural roads account for 40 percent of all vehicle miles traveled in this country. About 50 percent of rural roads are paved, and 90 percent of rural roads are one- or two-lane. There are 450,000 rural bridges. The rate of fatalities in rural areas is more than twice that of urban areas, mainly because the roads are narrower and people drive faster, resulting in more serious crashes.

Rural roads were originally designed so that they would fit the land the best way possible. They followed the ridges and the valleys and avoided steep slopes, wetlands, and other natural features that would have made construction more difficult. For the most part, these roads were designed for slower

Figure 7-1 The Big Sky Back Country Byway along Route 253 in Montana is in a rural area known for its sightseeing, wildlife-viewing, hiking, rock-hounding, bird-watching, hunting, and fishing. Image courtesy of Big Sky Back Country Byway.

speeds, so they may have tight curves, narrow lanes, narrow shoulders, and limited setbacks. Many of these roads still exist, although a number have been upgraded to meet current roadway design standards. Some of these incorporated green roadway standards, but most did not.

Many rural economies were built on a foundation of agriculture, mining, or forest products, and all three of these industries depend upon the highway network to transport their products. Many small rural communities were built with a main street and at a density in which walking is a very viable means of transportation. But over the years, they have grown to a point where walking is no longer an option.

A major problem in rural areas is that as growing metropolitan areas sprawl outward, agricultural uses and open space are typically lost, and impacts on cultural, natural, and scenic resources are often significant.

Types of Rural Development

There are many definitions of what "rural" means. The Rural Transportation Initiative defined three categories of rural areas and the roads typical to each:

Basic rural–Areas with few or no major population centers of at least five thousand people, and mainly agricultural in character. These areas often have declining populations or are growing very slowly. Most transportation projects are funded through federal programs.

Developed rural–Areas with one or more population centers of at least five thousand people, where the economy is a mix of agricultural, industrial, and service-based uses. These populations are typically stable, and transportation needs are greater to accommodate greater economic uses.

Urban boundary rural–Areas near metropolitan areas that are highly developed and often are experiencing rapid growth. Existing roads are being upgraded to meet increased demands, and new roads are added based upon the availability of funding. The potential conflict between rural character and sprawl is a major concern.

In most rural areas, there is a well-established network of local roads, arterials, and county or state highways. A problem, though, is that approximately 40 percent of county roads are inadequate for current travel, and nearly half the rural bridges longer

FIGURE 7-2 In rural areas, roads are an integral part of the landscape and have narrow travel lanes, shoulders, and rights-of-way. Photo by A. Crane, courtesy of National Scenic Byways Online (www.byways.org).

than twenty feet are currently structurally deficient. In addition, many of the nation's roads are failing, especially rural roads bordering Canada and Mexico, a result of increased truck traffic that resulted from the passage of the North American Free Trade Agreement in 1994. Such increased use underscores the continuing important role of rural America's highway system in the shipment of freight.

In May 1999, the U.S. Department of Transportation introduced the Rural Transportation Initiative, which has the following objectives:

- Improve transportation safety in rural areas to reduce the incidence and severity of accidents and their associated costs.
- Allow residents of rural areas and small communities access to the destinations and goods to attain their desired quality of life.
- Provide the transportation service that will afford rural areas and small communities the opportunity to reach their economic growth and trade potential.
- Enhance the social strength and cohesiveness of small communities and protect the natural environment of rural areas.
- Maintain the national security and border integrity necessary for the well-being of all Americans.

FIGURE 7-3 Almost all of New York State Route 22 is a two-lane rural road that passes through only small villages. Photo by Washington County Planning Department, courtesy of National Scenic Byways Online (www.byways.org).

Planning for Rural Areas

The development of rural transportation plans differs per state, but as a general rule, planning is handled at the local or regional level, with states providing a broader framework. In most small cities and counties, transportation projects are a component of the local capital improvement programs. Regional planning organizations are engaged in transportation planning. Some are advisory, but others have control over some of the transportation funds. Local and regional agencies tend to plan for smaller, more localized projects, but a state department of transportation (DOT) usually takes a systemwide approach.

The statewide planning process sets statewide transportation policy, and the planning documents produced cover at least a twenty-year horizon. A Statewide Transportation Improvement Program describes all capital and noncapital projects or phases of transportation project development that will use federal transportation dollars.

There is some public transportation in rural areas, provided by a variety of private sector or not-for-profit organizations and various public agencies. However,

Figure 7-4 Ocracoke's Howard Street, just off the Outer Banks Scenic Byway, is a deeply shaded narrow lane paved only with crushed oyster shells. Image courtesy of J. Sipes.

FIGURE 7-5 Gray rocks line the Old Frankfort Pike for miles in the horse-ranch country between Frankfort and Lexington, Kentucky. These walls were originally created from rocks removed from adjacent fields. Photo by A. Crane, courtesy of National Scenic Byways Online (www.byways.org).

it is becoming less and less financially viable. In 1965, more than 23,000 rural communities provided daily bus service, but today only around 4,500 communities provide that same level of service. Currently, the only supplier of rail service for most rural areas is the federally subsidized, for-profit Amtrak, which operates in forty-five states.

The Suburbs

Typically, suburbs are developed in ways that depend almost exclusively on automobiles as the primary form of transportation. Most suburbs were "greenfield" developments, meaning that they were constructed where fields and forests used to be. As a result, the environmental impact of suburban development has been significant.

Suburbs really took off after World War II when the automobile became affordable, there was a need to house returning veterans, and the expansion of the National Highway System made it easier for people to move away from the cities. Suburbs typically have a much lower density and smaller population than urban areas, but offer the same services, including schools, health care facilities, and public works. Many suburbs have more parks and recreation spaces than urban areas because of the available space. Residents of suburban areas typically commute to the cities for work, and most suburbs are designed so that

FIGURE 7-6 This view of Merriman Avenue, Stellenbosch, South Africa, was taken from a pedestrian bridge that spans the road. Image courtesy of Wikipedia Commons.

trips require the use of the automobile. The quality of life and economy in suburban America depends on an efficient, coordinated transportation system that maintains links between urban and suburban areas. But this isn't what has been built in most places.

The expansion of suburbia quickly led to suburban sprawl, which has resulted in increased traffic, longer commutes, more congestion, loss of farmland and open space, water and air pollution, destruction of wildlife habitat and species, and an overall reduction in quality of life. In recent years, there has been a lot of interest in revitalizing suburbs just as we have revitalized many urban areas. This revitalization involves changing land use patterns by introducing commercial and office uses that provide jobs and a walkable downtown, so suburbanites don't have to commute as much. The economic downturn that started in the late 2000s has slowed the development of new subdivisions and has shifted emphasis in many parts of the country to retrofitting existing suburbs. Many older suburbs are dated, and their infrastructure needs to be updated to accommodate current transportation demands.

Livable, Walkable Communities

A major point of emphasis for most urban, suburban, and rural communities is to decrease dependency on the automobile and increase walkability.

FIGURE 7-7 Planted medians, narrow travel lanes, and adjacent bike lanes help reduce the impact of roadways. Image courtesy of EDAW.

The American Association of State Highway and Transportation Officials' (AASHTO) book *The Road to Livability* (2010) cites a "livability" objective—to use transportation investments to improve the standard of living, the environment, and quality of life for all communities. AASHTO executive director John Horsley said, "If enhancing livability is the objective of transportation legislation or regulation, then it must work for those who live in rural Montana just as much as it would for those in downtown Portland. Equating livability only to riding transit, walking, and biking, limits its relevance and excludes a wide range of improvements and community needs."

AASHTO's *The Road to Livability* describes how a full range of transportation options, including improvements to roadways, transit, walking, and biking, can enhance community livability. The report also offers a baker's dozen of techniques state DOTs can use to improve the livability of their communities:

- Create good-paying jobs.
- Stimulate the broader economy.
- Invest in green projects.
- Revitalize a small town's main street.
- Transform urban streets into neighborhood centers.
- Preserve scenic country roads.
- Create smart transportation solutions for tight times.
- Enhance neighborhoods through the Transportation Enhancement program.
- Make design responsive to community needs.
- Integrate transportation and land use.
- Use scenic byways to attract tourists and support local economies.

FIGURE 7-8 Ladera Ranch is a four-thousand-acre community located in southern Orange County in California. Within the community are villages, which are made up of ten or more neighborhoods. Image courtesy of AECOM.

FIGURE 7-9 Ladera Ranch has several green features, such as roundabouts at small intersections, narrow street widths, and landscaped medians. Image courtesy of AECOM.

FIGURE 7-10 Smyrna Market Village just outside of Atlanta, Georgia, is a pedestrian-oriented urban village with emphasis on streetscapes, landscaping, and openness. Image courtesy of J. Sipes.

- Promote walking and biking.
- Support travel and tourism.

Suburban and Rural Case Studies

The Rain Gardens of Maplewood, Minnesota

Managing stormwater runoff is a major issue in the United States because poor management leads to contaminated, polluted water sources and flooding. Much of the stormwater infrastructure in this country is outdated, and many municipalities have put off upgrading and replacing it for so long that it has effectively become a crisis. The U.S. Environmental Protection Agency estimated that it will cost $276.8 billion between 2003 and 2023 to address the problem.

Fortunately for Maplewood, Minnesota, a small city located about ten miles northeast of downtown St. Paul, a group of visionaries decided in the mid-1990s that there was a better way to handle stormwater than using curbs, gutters, and underground pipes. The city decided that rain gardens would be a more environmentally friendly and affordable approach, and the first gardens were constructed in 1997. Today, more than 450 gardens have been installed along Maplewood's roads, and another thirty gardens have been constructed on city land. No matter where you go in the city, you will see rain gardens.

A rain garden is essentially a depression in the ground that is planted like a garden, and its primary purpose is to collect stormwater runoff and allow it to infiltrate into the soil. Rain gardens can significantly reduce the amount of stormwater that flows into

Figure 7-11 In many new suburban developments, such as this one in Mechanicsburg's Walden in central Pennsylvania, the emphasis is on creating a more pedestrian scale with less focus on roads and cars.

Figure 7-12 A rain garden can be placed almost anywhere, but is most useful if positioned to collect runoff, such as near a driveway, road, or sidewalk, or at the base of a slope. Image courtesy of Jamie Csizmadia.

sewers and drainage ditches, thereby reducing erosion and sedimentation problems.

There are two basic approaches to stormwater management: conveyance and infiltration. The conveyance approach, which typically involves curbs, gutters, and underground pipes, focuses on getting rid of the water as quickly as possible. The infiltration approach to stormwater management seeks to integrate water back into the landscape. Rain gardens are one type of infiltration approach.

In 1995, the City of Maplewood received funding from the state-funded Legislative Commission on Minnesota Resources, and some of this money was used to retrofit a two-block-long streetscape in an existing subdivision to repair roadways and improve stormwater management.

Many streets in Maplewood are lined with trees and few have sidewalks. The rain gardens along public roads take advantage of the underutilized spaces in public rights-of-way, and also encourage adjacent property owners to participate in the process. These gardens are almost always associated with road improvement projects.

The public rain gardens (not shown) are typically ten to twenty acres in size and are typically not associated with road projects. These gardens handle discharge from storm sewer systems.

Maplewood encourages homeowners, developers, and businesses to consider rain gardens and other best management practices for their site. Currently, an environmental utility fee is charged against all developed parcels, and these fees help finance stormwater facilities. Residents and businesses can apply for a credit reduction in their environmental utility fee if they use best management practices such as rain gardens, pervious driveways, or pervious parking lots.

One of the biggest surprises in Maplewood is how prevalent the rain gardens are. It seems that everywhere you look there are rain gardens. There are gardens in both upper- and lower-class neighborhoods, and in new developments as well as established neighborhoods.

FIGURE 7-13 There are a wide variety of plants and flowers available to choose from when planning a rain garden. Image courtesy of Jamie Csizmadia.

FIGURE 7-14 Rain gardens can be as shallow as four inches deep or as deep as twelve inches. Most rain gardens are simply depressions in the ground, with no need for special soil or pipes. Image courtesy of Jamie Csizmadia.

The City of Maplewood understood that for the rain garden program to be successful it had to be supported by the public. The city has developed an outreach effort to help citizens understand the benefits of rain gardens. Before installing a project, city staffers meet with neighborhood residents to talk about the benefits and to address potential concerns. A comprehensive educational package, including a fact sheet, was developed, and the city conducts rain garden construction workshops on a regular basis. They sponsor planting days and enlist neighborhood volunteers to coordinate and conduct the planting for rain garden projects.

Overall, the rain garden program is perceived by most to be a huge asset for Maplewood. There is no cost to the homeowner, and in addition to enhancing water quality and reducing urban pollution, the gardens also enhance the visual character of neighborhoods and attract a wider variety of birds and butterflies. Most rain gardens are built in the grass areas adjacent to the road, but in some locations roadway paving is removed to help direct stormwater and to provide space for a garden. When paving is dug up in these situations, the material is recycled and used as a base aggregate to help with drainage. The city estimates that a typical rain garden project costs 75–85 percent of a traditional curb and gutter project and reduces stormwater runoff by as much as 80 percent annually.

State Route 17 in Horseheads, New York

State Route 17 runs across the state of New York, connecting Pennsylvania and New Jersey. In the town of Horseheads, New York, Route 17 was a four-lane highway with both grade-separated and at-grade interchanges. In 1996, the New York State DOT completed a formal scoping study and determined that either the reconstruction or the realignment of Route 17 was necessary.

Once a study of alternatives was completed, a design was chosen that minimized negative impacts while improving safety and the flow of traffic in the area. The selected alternative—an elevated expressway that follows the existing alignment—has proven to be cost-effective, involve fewer environmental impacts, and enable businesses to stay open during construction. Bridges were constructed at three locations to allow for north-south traffic. Local traffic was accommodated with one-way service roads along the north and south sides of the elevated expressway. A linear park running parallel to the highway serves as a buffer for residences and as a community enhancement.

The design of the preferred alternative was completed in early 2004, and construction began in April 2004. The $60-million project was completed in September 2007 (Struve and Breen, 2009).

High Point Redevelopment, Seattle, Washington

High Point is a mixed-income, mixed-ethnicity planned community in West Seattle that is built around concepts of New Urbanism. It includes new housing, new streets that are walkable, and parks and amenities that help create a sense of place. Another key driver of the design is a strong commitment to energy efficiency and green building practices.

Figure 7-15 The redevelopment of High Point in Seattle features the only porous pavement in Washington State. Image courtesy of Wikipedia Commons.

Figure 7-16 One example of a sustainable feature in High Point's redeveloped community is the use of rain gardens. This design allows more stormwater to soak into the ground, thereby reducing pollution. Image courtesy of Wikipedia Commons.

The original High Point community was a government-funded low-income development built as military housing in the 1940s in the wake of World War II. Over the years the neighborhood began to deteriorate, and by the 1980s, High Point was a dangerous place to live; drugs and gangs were prevalent throughout the neighborhood.

In 2001, a group of residents began to publicize a vision of High Point as a mixed-income, mixed-ethnicity community, with new streets, parks, and amenities knitted more closely into the fabric of the larger community. The project is the Seattle Housing Authority's largest family community, and initial funding came from a $35 million HOPE VI grant awarded by the U.S. Department of Housing and Urban Development in 1999. The HOPE VI program funded projects focused on correcting extremely distressed public housing. The total funding for the project, including public and private sources, totals more than $550 million. Low-income housing was integrated with market-value homes, and other facilities included parks, a community center, a library, and a dental clinic/health clinic.

Public involvement is of the utmost importance when it comes to redeveloping a community. From interactions with the community, the design team was able to define project goals. One of these goals was to correct the existing street system. The existing circulation pattern was not only confusing, but it was also largely disconnected from streets in surrounding neighborhoods.

New streets were designed to be more livable and were realigned and reconnected to the West Seattle grid. Today most of the streets in High Point are fairly narrow and have parking only on one side of the road. This approach helps slow down traffic because oncoming cars must wait and allow each other to pass. It also results in less paving. To further calm traffic,

traffic circles are used at many intersections. Planting strips along streets are wide to accommodate street trees.

One major objective of the High Point Redevelopment project was to improve the quality of water of the existing watershed that runs into Longfellow Creek. This is one of the three largest natural streams in Seattle, and is the city's most significant salmon-producing stream. Instead of using typical curb and gutter solutions, natural swells are created between the streets and the sidewalks. The roadway is slanted toward the swell.

All the sidewalks in the development are made of porous concrete, which allows the water to seep through the concrete and into the ground. The rainwater will soak into the ground, either through the sidewalk or into the swell. Once the ground is saturated, the water will flow along the swell and into Longfellow Creek. The water is being filtered by gravel and stones. A large detention pond has been designed as an attraction. A walking trail surrounds the pond, and an artificial stream, complete with boulders and a bridge, has been built.

Chapter 8

Cultural/Historic/Visual Resources

What would be the point of building roads if there were no people? Roads exist only because we need them to get from one place to another. We need to start thinking of roads as a critical part of quality of life. Within that context, roads should be constructed or reconstructed so that they are respectful of cultural, historic, and visual resources. Green roadways are as much a part of a community as parks, schools, churches, office buildings, shopping malls, and residential structures.

Cultural and Historic Resources

Cultural resources are considered to be the collective evidence of the past activities and accomplishments of people. They include prehistoric and historic archaeological sites, historic standing structures, bridges, cemeteries, and monuments, among others. All federal and most state agencies are required to identify and protect cultural resources on the lands they manage. The concerns of Native Americans are also addressed by a variety of specialized federal and state laws and regulations.

Congress has made historic preservation a responsibility of every federal agency, including the Federal Highway Administration (FHWA). The two primary laws that apply to transportation projects and their impacts to cultural resources are Section 4(f) of the Department of Transportation Act and Section 106 of the National Historic Preservation Act. (See chapter 2 for more information on Section 4[f]).

Section 106 of the act is the basic federal law requiring all federal agencies to take into account the effects of their actions on historic properties. The key participants in the Section 106 process are the state historic preservation officer, the tribal historic preservation officer for projects on reservations, and the Federal Advisory Council on Historic Preservation, which is an independent agency created by Congress. This agency is responsible for identifying historic properties that may be eligible for listing in the National Register of Historic Places.

The purpose of cultural resource investigations is to consider the impact of federally funded undertakings on properties, sites, buildings, structures, and objects that are listed in or may be eligible for inclusion in the National Register of Historic Places.

Much can be done to create awareness about the road's heritage, attract tourism, and enhance the economic climate of the towns along the byway. In the United States, travel and tourism generated $1.9 trillion in economic output, with $812.9 billion spent directly by domestic and international travelers in 2011.

Figure 8-1 Historic structures, such as this old covered bridge, are an important part of the existing culture and should be protected. Image acquired from depositphotos.com.

Figure 8-2 The first planned scenic highway in the United States was the Historic Columbia River Highway. This seventy-five-mile highway was constructed in Oregon between 1913 and 1922. Photo by Dennis Adams, courtesy of National Scenic Byways Online (www.byways.org).

FIGURE 8-3 Bright white wooden railings, part of the Historic Columbia River Highway's cultural heritage, line a strip of the highway. Photo by Dennis Adams, courtesy of National Scenic Byways Online (www.byways.org).

The FHWA Historic Preservation and Archeology Program provides guidance and technical assistance to federal, state, and local governments regarding historic preservation and cultural resources. In 1999, the Transportation Research Board's Historic and Archaeological Preservation in Transportation Committee, with assistance from FHWA and the National Park Service, organized the National Forum on Assessing Historic Significance for Transportation Programs. Each state also has its own guidelines for protecting cultural and historic resources, which are typically available on the state's Department of Transportation (DOT) website.

Some recommendations for protecting cultural and historic resources are:

- Revitalize those properties or features that do not currently contribute in a positive way to the scenic, cultural, or historic character of the road.
- Maintain individual historic buildings, towns, or roadway elements for the future by following good maintenance and rehabilitation practices.
- Interpret the history of the road through signage and place names that identify significant features.
- Develop pull-offs at historic locations to allow tourists or visitors the opportunity to obtain interpretive information about a road and surrounding features.
- Preserve and protect those characteristics of the scenic byway that contribute in a positive way to its scenic, cultural, and historic character.
- Reduce sign clutter as well as the visual impact of the signs on surrounding areas.

Figure 8-4 Gettysburg National Military Park is dedicated to preserving and understanding the heritage and significance of Gettysburg. The park has many symbolic monuments. Photograph by Gettysburg National Military Park, courtesy of National Scenic Byways Online (www.byways.org).

Figure 8-5 A young horseback rider explores Smugglers Notch Scenic Byway, a scenic road through a rocky, alpine landscape in Vermont. Photo by A. Crane, courtesy of National Scenic Byways Online (www.byways.org).

Archaeology

Archaeological studies are required for virtually any major transportation project. Early on in the National Environmental Policy Act (NEPA) process, the decision is made whether there are any significant findings or problems associated with archaeology. The survey of a property for archaeological resources may consist of an in-house review or a field survey, or both. A reconnaissance survey, or Phase I survey, characterizes the archaeology of a region and identifies areas where archaeological sites are likely to be discovered.

When an archaeological survey fails to locate any archaeological resources within the area of potential environmental effect of the proposed project, the finding of "no archaeological resources" is determined. If archaeological resources are found, an intensive survey, or Phase II survey, must be conducted. Phase II archaeological research entails more detailed background studies, historical research, field investigations, and analysis than Phase I surveys. A Phase II study also determines if a site is eligible for the National Register of Historic Places. A Phase III study involves the recovery of archaeological resources to mitigate adverse impacts.

Visual Resources

One of the basic concepts behind the idea of green roadways is to protect visual resources. If a road project is to be constructed, it should be done so as to minimize negative impacts while visually blending with the surrounding landscape and community.

Both the U.S. Environmental Protection Agency (EPA) and FHWA have recognized that visual, scenic,

FIGURE 8-6 The Gaoliang Bridge in China was built to honor a brave warrior from ancient Chinese legends. Image courtesy of Wikipedia Commons.

and aesthetic qualities of a landscape are important parts of a transportation project. Regardless of the land use, the objective is to achieve an appropriate visual fit between a transportation facility and its surroundings, regardless of whether it is in an urban, suburban, or rural setting. The impact of a roadway in a rural landscape can be more pronounced than in urban settings because the visual contrast is so significant and natural environments are disturbed. In contrast, urban settings are dominated by structures connected by a network of transportation links and utilities. The aggregation of buildings, streets, signs, power distribution lines, light standards, and the like, combines to create a very complex visual environment in which the creation of a road is a less stark change.

Viewsheds

The visual impact of a roadway goes far beyond the right-of-way. A good rule of thumb for every project where visual resources may be an issue is to determine the viewshed for a given roadway. A viewshed is basically every area from which the roadway would be visible, and is typically modeled to show both the view *from* the road as well as the view *of* the road. In hilly areas with dense vegetation, the viewshed may be quite small because visibility is limited. In areas with flat or rolling terrain, a roadway may be visible for miles.

Viewsheds are determined by analyzing digital elevation models using geographic information system software. All areas that are visible from the highway are combined to create the viewshed.

The viewshed in an urban landscape tends to be limited. Panoramic views in urban settings are most likely associated with high bridge structures or roads that ascend major topographic features. Criteria that may be used to select a key view location include visibility of the project area from the viewpoint, frequency and duration of the public viewing time, and the similarity of the view to a larger portion of the project.

Negative visual impacts to viewers of the road are much longer lasting than views from the road, so they are given much greater consideration. In residential areas it is usually desirable to restrict views to and from the highway corridor. However, in commercial, institutional, and industrial zones, maintaining views to and from properties becomes very important.

Approaches for Addressing Visual Resources

NEPA requires that visual resources be considered during the assessment of a transportation project, including both the view of the road and the view from the road. Additional significant federal legislation that focuses on visual resources includes the Highway Beautification Act of 1965, which for the first time set aside federal funds to be used specifically for programs that limited outdoor advertising or removed or screened "offensive" uses such as junkyards along highways, and for scenic enhancements (FHWA, 1976: 369).

Each state has its own approach for addressing aesthetics. Some states take a general approach. Some are developing detailed design standards, and others focus on design solutions at the local level. Some states have developed a more comprehensive approach, including the following:

Arizona has implemented landscape design guidelines for urban areas, freeway mitigation and enhancement, erosion and pollution control, integrated natural resource management, and urban forestry. Some cities require a more detailed look at highway aesthetics; for example, Scottsdale advocates planning and design in accord with "character areas," to maintain visual character of the city.

California has developed and organized a system for planning and designing landscape and aesthetic improvements. The California Department of Transportation, typically referred to as Caltrans, has implemented programs that create context-sensitive highway designs, use native plant materials, incorporate transportation art and aesthetics into highway structures, and ensure consideration of community values along with safety, economics, and mobility. The state's *Highway Design Manual* includes sections on landscape and aesthetics, and the landscape architecture program provides direction and coordination for context-sensitive solutions; training development; erosion control and highway planting policies, standards, and guidelines; landscaped freeway designations; roadside management; and research and new technology.

Michigan DOT's *Aesthetic Project Opportunities Inventory* lists approximately two thousand opportunities for improving the visual quality of the environment along state highways. The inventory identifies eight types of aesthetic projects: landscape treatment opportunities, streetscaping opportunities, site or corridor management plans, scenic easement acquisitions, scenic turnout sites, structure removal or improvements, vegetation management opportunities, and landform improvements.

Minnesota's *Highway Project Development Process* provides technical guidance on subjects such as vegetation, visual quality, noise, and soils. Another document defines the state's vision for addressing landscaping and aesthetics through case studies from the past twenty-five years and describes ten characteristics that contribute to noteworthy environmental effects.

New Jersey's *Landscape and Urban Design Unit Procedure Manual* integrates landscape and aesthetics with highway design, community participation, and construction. The manual includes checklists and forms for plan reviews, landscape design, soil erosion and sedimentation, noise barrier aesthetics, wetlands design, final plan review, and monitoring scenic lands.

Ohio DOT's design standards and guidelines incorporate patterns, colors, texture, and landscaping to increase the visual appeal of highways, noise barriers, and bridges for motorists and residents. The agency estimates that the cost for improved aesthetics amounts to less than 1 percent of a project's total cost. Ohio DOT's Gateway Landscaping Program helps towns and cities improve landscaping along the highways leading into their communities.

Texas DOT addresses the visual characteristics of highways in the *Landscape and Aesthetics Design Manual*, which describes aesthetic approaches for highway design and provides general guidance on the applications. A supplement, *Develop Cost-Effective Plans to Add Aesthetically Pleasing Features to Transportation Projects*, guides Texas DOT designers and consultants in developing and constructing aesthetic treatments (Sipes and Blakemore, 2007).

U.S. Forest Service Visual Management System

The Visual Management System (VMS), developed by the U.S. Forest Service (USFS) in 1974, set the standard for addressing landscape aesthetics. The VMS included objective criteria such as distance of view and visual magnitude, and it is based on subjective definitions of aesthetic landscapes and expert assessment of impacts based on classical principles of art and beauty. In the years since, understanding of ecological succession, disturbance agents, and effects of various management practices has increased. In 1996, the USFS published "Landscape Aesthetics: A Handbook of Scenery Management," which defined the basics of a scenery management system. This approach builds upon the foundations of the VMS, and takes a more integrated approach that includes broader land and resource management systems and processes of the agency.

In 1981, FHWA developed the Visual Impact Assessment Methodology for Highway Projects to provide

Figure 8-7 Thick forests of bright yellows and dark greens nearly hide West Elk Loop, Colorado, in autumn as it curves through stands of trees. Photo by Rob Strickland, courtesy of National Scenic Byways Online (www.byways.org).

guidance in analyzing and quantifying visual impacts for highway proposals. It remains the standard methodology for this purpose. In recent years, some DOTs have modified this methodology to meet their needs.

Visual Assessment Process

Several states have developed a visual assessment process that follows guidelines outlined in FHWA's *Visual Impact Assessment for Highway Projects* and identifies seven principal steps: (1) define project setting, (2) define viewsheds and key views, (3) determine visual quality, (4) determine viewer sensitivity and exposure, (5) depict visual appearance of project alternatives, (6) conduct visual impact assessment, and (7) mitigate adverse visual impacts.

Visual Resource Recommendations

The visual design elements of material, color, texture, pattern, and form collectively establish the architectural and visual framework of transportation infrastructure and most directly affect perceptions of the infrastructure and the surrounding environment. These design elements can be applied in a wide variety of ways to address aesthetics.

Figure 8-8 Haines Highway in Alaska was built in 1943 as an alternate route from the Pacific Ocean to the Alaska Highway in case the White Pass & Yukon Route Railroad should be blocked. Photo by Lori Stepansky, courtesy of National Scenic Byways Online (www.byways.org).

Figure 8-9 The Cape Horn Overlook on the Columbia River Gorge along Washington State's Highway 14 provides a breathtaking view. Photo by Dennis Adams, courtesy of National Scenic Byways Online (www.byways.org).

FIGURE 8-10 Rafting is a popular recreational activity along the Ocoee River, which runs through Tennessee's Cherokee National Forest, located in the southern Appalachian Mountains. Image courtesy of U.S. Forest Service, Cherokee National Forest.

Design Approaches

- Employ materials similar to those in the adjacent landscape. This is particularly important in urban centers where the built landscape is dominant.
- Use similar colors. Quite often it is impractical to use or attempt to match the materials of the adjacent landscape. In these cases color becomes the single most important tool.
- Use similar plant materials to blend the landscape. Use native, informal plantings along the roadway and more formal/ornamental plantings in urban areas.

In the rural setting landscape materials can supplement and link existing landscape features.

Road Layout

- Design road lines to curve gently and blend with the landform by climbing in hollows and dropping on ridge lines.
- Align roads diagonally to slopes in those situations where midslope roads cannot be avoided; vary alignment in response to landform.
- Curve road lines gently to blend with natural landforms, dropping on convex slopes and rising in hollows.
- Avoid locating roads that follow the viewers' line of sight on gentle foreground slopes; roads should curve away or cross at another angle.

Figure 8-11 A stone retaining wall along the side of the road minimizes impacts to adjacent natural resources. Image acquired from depositphotos.com.

Landform

- Be sensitive to the visual character of the landform. Landform can be a dominant element of the roadway, particularly in hilly or mountainous terrain. Exposed rock faces, steep cut slopes, and high fills can be dramatic in scale but are often objectionable if they bisect existing landscape features considered visually pleasant or socially significant.
- Rock features should emulate outcrops and engineered features should reflect the use of native bedrock.
- Design road locations to make as much use of landform as possible and take advantage of nonvisible areas, benches, and vegetative screening wherever possible to reduce visual effects.
- Reduce the size of cut-and-fill slopes to decrease contrast between a road and the landscape.

Scenic Byways

In 1991, the federal government created the National Scenic Byways Program, to protect and beautify

Figure 8-12 A scenic viewing area on the Klehini River in Alaska provides an interpretive sign that gives information about the local ecosystem. Photo by Lori Stepansky, courtesy of National Scenic Byways Online (www.byways.org).

FIGURE 8-13 A highway sign notifies travelers that they are driving along the Rogue Umpqua Scenic Byway. Located deep in the Cascade Mountains in Oregon, the byway is most commonly known as the "highway of waterfalls." Photo by John Sload, courtesy of National Scenic Byways Online (www.byways.org).

roadways noted for their scenic, natural, historic, cultural, archaeological, or recreational quality.

According to the FHWA, the designation as a scenic roadway has four significant benefits: preservation, promotion, pride, and partnerships. Preservation of vistas, roadside scenery, and historic buildings can be facilitated through the program. The Highway Beautification Act of 1965 prohibits new billboards along designated scenic byways that are interstate, part of the National Highway System, or federally aided primary roads.

To meet the goal of visual harmony, the design characteristics of individual elements must be coordinated to provide a clear sense of order, clarity, and continuity. The visual design elements of material, color, texture, pattern, and form all influence the final decisions for a road. The result is not only a green roadway that respects the existing visual character, but also enhances the driving experience.

The FHWA manages the National Scenic Byways Program, for which Congress provided more than $380 million for 2,832 state and nationally designated projects as of 2010. The program is a grassroots collaborative effort established to help recognize, preserve, and enhance selected roads throughout the United States.

The U.S. Secretary of Transportation recognizes certain roads as All-American Roads or National

FIGURE 8-14 The stone-arched bridges of the Blue Ridge Parkway blend into the background, adding character and charm to the scenery. Image courtesy of Wikipedia Commons.

Figure 8-15 Built in 1909, Campbell's Covered Bridge is the last remaining covered bridge in South Carolina. Photo by Dennis Egan, courtesy of National Scenic Byways Online (www.byways.org).

Scenic Byways based on one or more archeological, cultural, historic, natural, recreational, or scenic quality. All-American Roads are the very best of the National Scenic Byways. For example, the San Juan Skyway, in southern Colorado, is a 233-mile scenic loop road that weaves through the San Juan Mountains and links the towns of Ridgway, Durango, and Cortez. The views from the skyway are incredible, and given the steep slopes, sheer cliffs, and sharp switchbacks of the area, it clearly took great consideration to integrate the road and the landscape.

Figure 8-16 An orientation kiosk is positioned at the north gateway of Smugglers Notch Scenic Byway as a guide for visitors. Photo by A. Crane, courtesy of National Scenic Byways Online (www.byways.org).

FIGURE 8-17 The Florida Black Bear Scenic Byway is a 123-mile highway that passes through Ocala National Forest. The forest is home to rare and endangered plants and animals that cannot be found anywhere else. Photo by U.S. Forest Service, courtesy of National Scenic Byways Online (www.byways.org).

Case Studies—Cultural/Historic/ Visual Resources in Transportation

I-70 Glenwood Canyon and Snowmass Canyon, Colorado

The Rocky Mountains in Colorado include some of the most beautiful and rugged areas in the United States. This is also one of the most difficult places to build a highway while also respecting picturesque mountains and lavish forests that are home to an abundance of wildlife. Glenwood Canyon and Snowmass Canyon, although they are different projects, are very similar in many ways. They both contained two-lane roads that no longer met transportation demands, so plans were made to update and widen each. Both also were constructed along steep cliffs, so room for expansion was limited.

Glenwood Canyon was the first of the two projects to be constructed. The road was widened from two to four lanes, and many horizontal and vertical deficiencies were corrected. The new roadway has four full-service rest areas, a bike path running the entire length of the roadway, and designated pull-outs for launching boats or kayaks. The roadway also includes forty bridges and viaducts stretching over six miles, fifteen miles of retaining wall, and almost a mile of tunnels in both directions. Construction on the project began in 1980, and the road was opened to traffic in 1992. The 12.5-mile road completes the final leg of I-70.

Once Glenwood Canyon had been built, the planning for Snowmass Canyon began. Snowmass is only thirty-five miles from Glenwood Canyon, and the projects are remarkably similar. Snowmass contained a two-lane road that needed to be widened to four lanes, and it had limited right-of-way for expansion. Improvements to the existing road were important because it was the final piece to State Highway 82.

Snowmass Canyon was considered to be the more complicated of the two projects. The geotechnical studies of the area alone took more than two years to complete. In the end, the Colorado DOT decided to employ numerous strategies to stabilize the hillside; they used many types of tiebacks and retaining walls, some as long as four thousand feet. Trees were also avoided as much as possible. Disturbed areas were replanted with native vegetation. The road through Snowmass Canyon was finished in 2004, one year ahead of schedule.

The roads in Glenwood Canyon and Snowmass Canyon are amazing works of engineering. Widening a narrow roadway in the heart of three hundred–foot canyons is challenging in itself. Engineers found creative ways in both canyons to minimize the footprint of the roadway while maintaining the landscape of Colorado.

Paris Pike, Kentucky

In the center of Kentucky, just outside of Lexington, is a small town called Paris, and the road from Paris to

FIGURE 8-18 Approaching the entrance to the Glenwood Hanging Tunnel along I-70 in Colorado, viaducts take I-70 directly into the tunnel portals. Image courtesy of Wikipedia Commons.

FIGURE 8-19 The westbound road to Hanging Lake Tunnel, along Glenwood Canyon, is positioned on a high viaduct, providing travelers with a good view. Image courtesy of Wikipedia Commons.

Lexington is called Paris Pike. The road was a beautiful two-lane country road traversing through the rolling, scenic horse farms of rural Kentucky.

Unfortunately, Paris Pike was also one of the most dangerous roads in the state. The travel lanes were too narrow, there were no shoulders or passing lanes, the curves were too sharp, and the horizontal and vertical curves were not synced. The horizontal alignment was straight, and motorists felt like they could see a long distance. The problem, though, is that the vertical alignment changed significantly. The result was that motorists would move to pass because they didn't see any other cars, but in reality there could be a car in a low spot of the road that was not visible.

In the early 1970s, the Kentucky Transportation Cabinet (KTC) originally decided that the road needed to be expanded from two lanes to four to address both safety problems and level of service. For twenty-seven years, citizens knew the road needed to be fixed, but local concerns about preserving the natural environment and history prevented the project from proceeding. The problem was the limited available space for expansion within the existing right-of-way. Large, mature trees, hundred-year-old dry-stack stone walls, wooden slat fences, historic houses, and rolling pastures along both sides of the roadway would be affected by a road expansion. Surrounding landowners and environmental groups fought to preserve the character of the landscape and were opposed to any project. However, when a local family was killed on the road in 1985 during the Christmas holiday, the state decided something had to be done.

KTC, key stakeholders, and the local community worked together to reach a compromise. KTC hired Jones & Jones, a landscape architecture firm from Seattle, Washington, to work with H. W. Lochner, a local engineering firm, to preserve the natural beauty of the road. After much deliberation, KTC decided to widen a 13.5-mile stretch of the road from a two-lane road to a four-lane road using context-sensitive solutions. Instead of simply widening the existing road, the road was built as two separate roads—one northbound and one southbound—with a wide median between the two. Grant Jones, founder of Jones & Jones, referred to this as the "land between the roads," and the basic idea was to create the feeling of driving on a rural Kentucky road. These new roads followed the contours of the land and wove through the existing landscape, preserving existing cultural and natural features. Jones & Jones designated "zones of opportunity," or areas that could be widened without encroaching on sensitive areas. Where existing trees, fences, or stone walls were affected, they were either moved and reconstructed somewhere else to their original condition, or replaced with similar features.

One reason that Paris Pike is a success is that the natural beauty and cultural importance of the existing landscape was considered when designing the road. The shoulders of the road are grass reinforced with geotextiles so they can support cars, and guardrails are constructed of wood backed by steel so they not only keep the visual character of "horse country," but meet

FIGURE 8-20 Care was taken during the widening project of Paris Pike in 1997 to preserve the existing rock walls and entryways and to let the highway flow with the contours of the landscape. Image courtesy of Jones & Jones.

safety standards, too. The median preserved many of the mature trees that define the character of the area, and the road was aligned to minimize impacts. Paris Pike even has a visitor center that is housed in an historic farmhouse along the roadway.

The new four-lane Paris Pike is integrated into the Kentucky landscape. It weaves through the beautiful horse farms, around historic fences dating back to the Civil War, and past hundred-year-old trees, and the best part is that motorists can use the roadway safely.

Old Florida Heritage Highway, Florida

The Old Florida Heritage Highway (OFHH) corridor travels through several historic communities (Micanopy, Evinston, Cross Creek, Rochelle, and Island Grove) and traverses Paynes Prairie, a national natural landmark and Florida's first state preserve. The fifty-mile-long OFHH received its official designation in June 2001.

The unique corridors in Alachua County and their associated cultural and natural resources are some of the most compellingly beautiful landscapes that this part of Florida has to offer. Like many rural areas in the southeast, southern Alachua County is experiencing uncontrolled sprawl that is spreading out from the neighboring urban areas and is threatening the visual, cultural, and environmental resources of the area.

The master plan for the OFHH corridor is intended to establish the value and importance of scenic corridors and to define how they shape and influence communities along the highway. Conceived of as a catalyst for the continued development of heritage and nature tourism, the master plan, to be successful, must help create the motives and mechanisms for preserving the character and integrity of the region's rich natural and cultural resources.

The design guidance provided by the master plan, although conceptual in nature, is inspired by the uniqueness of Alachua County and its people. The design themes and standards for the recommended improvements are drawn from a thorough and careful understanding of the land, the community, and the history that has shaped both.

The plan for the OFHH focused on three design types: corridors, nodes, and trails. The corridors are linear in nature and include the areas adjacent to roads, water bodies, and existing trails. Along roads, the corridor includes not only the right-of-way, but

FIGURE 8-21 An aerial view of Paris Pike shows how the north- and southbound lanes move independently of each other, creating a median with variable widths. Image courtesy of J. Sipes.

FIGURE 8-22 The unique landscape of the Old Florida Heritage Highway is enhanced with lakes, prairies, wetlands, mature live oaks, and rural homesteads. Image courtesy of J. Sipes.

also adjacent areas that are part of the viewshed and have a significant impact on visual character. Design recommendations for corridors vary depending upon each specific corridor. Within each corridor, design recommendations focus on what happens within the right-of-way because that is under the control of the county. The key is how to meet transportation demands, address safety and access, and enhance aesthetics and nonmotorized use.

Nodes are specific sites located throughout the study area where people may want to stop. From a design standpoint, these are gateways, visitor centers, interpretive centers, kiosks, and wayside signage. Nodes give motorists an opportunity to stop, get out of their cars, and learn more about the OFHH. Each node includes interpretive information and appropriate maps to orient visitors and let them know of other opportunities in the area. Some nodes are also geared more toward pedestrians, cyclists, and equestrians.

Different alternatives for the OFHH were originally based on two basic design themes. One set of alternatives focused on "settlement" and the other was oriented more toward "nature." The final alternative concepts often consisted of a blending of the two themes.

A basic idea behind all design concepts is that they build upon the vernacular of the region. That means using colors, shapes, textures, and materials that are consistent with what can be seen in the area. The basic forms for structural elements are based on the roof shapes found in the homes, barns, and other structures that occur along the corridor. The various design components vary in size and form depending upon location. For example, in some locations, kiosks are primarily oriented for cyclists, but in others they are geared toward motorists driving along the corridor.

Next Steps

The next step in the planning process was to identify high priority items as part of a viable implementation strategy. For the OFHH to be successful, it needed to clearly define an identity embraced by both local residents and visitors.

One of the major reasons the OFHH has been successful is that numerous individuals, organizations, and nonprofits have partnered with Alachua County to help protect the character of the corridors. The OFHH played an important role in facilitating coordination in its work with these diverse groups. In particular, the acquisition of land for preservation purposes was approached as part of a larger organized effort with clearly defined goals and objectives. A maintenance agreement was developed among the county, Florida DOT, and/or other public or private agency partners to address how landscaping above Florida DOT minimum levels will be cared for within the OFHH right-of-way.

Merritt Parkway, Connecticut

The Merritt Parkway, a thirty-seven-mile, four-lane divided roadway located in Fairfield County, Connecticut, is one of the oldest paved parkways in the country. It is considered by many to be one of the nation's best examples of a green roadway because of the preservation of the surrounding landscape and the enhanced driving experience for motorists. It has often been called "The Queen of Parkways."

Figure 8-23 The Merritt Parkway shield welcomes motorists to a scenic, historic parkway in Fairfield County, Connecticut. Image courtesy of Wikipedia Commons.

The parkway was designed as a limited-access road in large part because of concerns that New Yorkers would overrun the countryside. Construction began in the 1930s, and the entire parkway was opened to traffic in 1940. The first 17.5-mile segment of the parkway, from the state line to the town of Norfolk, cost $20 million to construct. Tolls were added in 1939 to generate funding for an expansion of the parkway. Originally, each motorist had to pay a dime at a toll booth, but the fee increased over time. The tolls were finally removed in 1988 in large part because of safety concerns.

Timelessness

The parkway is one of those places that is reminiscent of a quieter, more gentle time in this country's history. This was a time when America discovered the automobile and there was a fascination for the concept of "driving for pleasure."

A basic concept behind the Merritt Parkway is to preserve the existing character of the landscape and create a pleasurable driving experience. As its name suggests, the experience is similar to driving through a park with the large deciduous trees, flowering trees and shrubs, open meadows, and rolling hills that make up the context for the parkway, which was designed at a time when the emphasis was on driving for pleasure. The speed limit on the parkway still ranges from forty-five to fifty-five miles per hour. This lower speed limit allowed the designers to focus more on aesthetics and the overall driving experience than is typically done for projects today. There are reports of motorists stopping for picnics along the side of the road.

Once completed, the Merritt Parkway quickly became a national model for how to build roads. For many, it is a cultural icon that represents a simpler time and is representative of Connecticut's heritage. The parkway was placed on the National Register of Historic Places in 1991, and in 1993 was designated a State Scenic Road. In 1996, it was designated a National Scenic Byway.

The Merritt Parkway is specifically known for its diverse collection of Art Deco, Gothic, French Renaissance, and Art Moderne bridges. There were originally sixty-nine bridges, designed by architect George L. Dunkelberger.

The basic function of the parkway has not changed significantly over the years, although it handles a greater volume traffic at much higher speeds than was originally intended. This has led to concerns about safety. Trees are very closer to the edge of the road and both shoulders and median are very narrow. This combination has led to a high number of traffic accidents.

FIGURE 8-24 During construction of the Merritt Parkway in the 1930s, landscape architects attempted to save as many trees as possible from being cut down by creating gently rounding hills along the highway. Image courtesy of Connecticut Department of Transportation.

FIGURE 8-25 The Frenchtown Road Bridge is a concrete arch bridge that was built in 1942 over the Merritt Parkway. Image courtesy of the Library of Congress-HABS/HAER.

The Merritt Parkway has been affected by extensive growth in Fairfield County, and rush-hour traffic can be a nightmare. It is still a major commuter thoroughfare for motorists traveling between New York and Connecticut. It has been difficult to maintain the original character and intent of the parkway while meeting current transportation requirements. This has led to battles over the years about what improvements, if any, were needed for the Merritt Parkway.

The state has constructed several interchange improvements that included lengthening merges, smoothing out ramp curves, and removing ramps to simplify access. Signage along the parkway was updated in 2001. Vistas have been cut through the surrounding forest and dense undergrowth to provide glimpses of the countryside beyond. Several new expressways built over the years cross the Merritt Parkway, and this has changed the overall look and feel of the landscape because the new interchanges are much larger than those of the original parkway. A better solution would have been to plan the expressways so they didn't have to cross the parkway.

There were discussions in the 1980s to widen the parkway to eight lanes, and some alternatives looked at straightening the curves to increase the speed limit. In 2002 and 2003, there were discussions about widening the parkway to six lanes, but the state Transportation Strategy Board rejected the proposal because of public resistance.

FIGURE 8-26 The roadway is designed to enhance the driving experience while preserving existing natural and cultural resources. Image courtesy of Wikipedia Commons.

Even though the changes to the parkway have been managed pretty well so far, there is concern about its long-term health. In 2010, the National Trust for Historic Preservation identified the Merritt Parkway as one of the eleven most endangered historic areas in the United States because of neglect and poor planning. Lack of funding has led to inadequate maintenance, and some of the original bridges have been in disrepair. Over the years, several of the bridges have been torn down and replaced.

Despite concerns, though, the Merritt Parkway continues to be an example of how to make transportation projects attractive and memorable. There are still no billboards or trucks allowed on the parkway, and if you can avoid the traffic, the drive is still one of the most pleasurable in the country. One reason that the parkway has maintained its parklike setting despite encroaching development is that the roadway was built in the northern half of a three-hundred–foot right-of-way. This was originally done to control access to the road and allow for future widening, but it also helped maintain the country character of the road.

Nevada Landscape and Aesthetics Master Plan

When it came to designing roads, the Nevada DOT historically took a utilitarian approach emphasizing safety and operations, with little concern for cultural, natural, and aesthetic resources. DOT staff followed that same philosophy when they initiated work on the Carson City Bypass in the late 1990s. The process was going okay until designs for the bypass were 60 percent complete, and at that point several organizations and citizens in Carson City banded together to stop the project because included freestanding walls that would block views of the surrounding mountains did not remedy the split already felt in the community, and bridges and other improvements lacked visual quality. The DOT had to abandon the original plans and go back to square one, costing Nevada tax payers millions of dollars. Nevada's governor at the time, Kenny C. Guinn, vowed that the state would not be put in that position again, and he directed the DOT to develop a strategy to help avoid similar problems in future transportation projects.

Developing the Master Plan

In May 2000, the Nevada State Transportation Board and Nevada DOT initiated a process to create a statewide master plan for landscape and aesthetics. The agencies contracted with the landscape architecture program at the University of Nevada, Las Vegas, which had been working across the state on community-oriented planning projects. "The original concept was to design every highway in the state, but it did not take

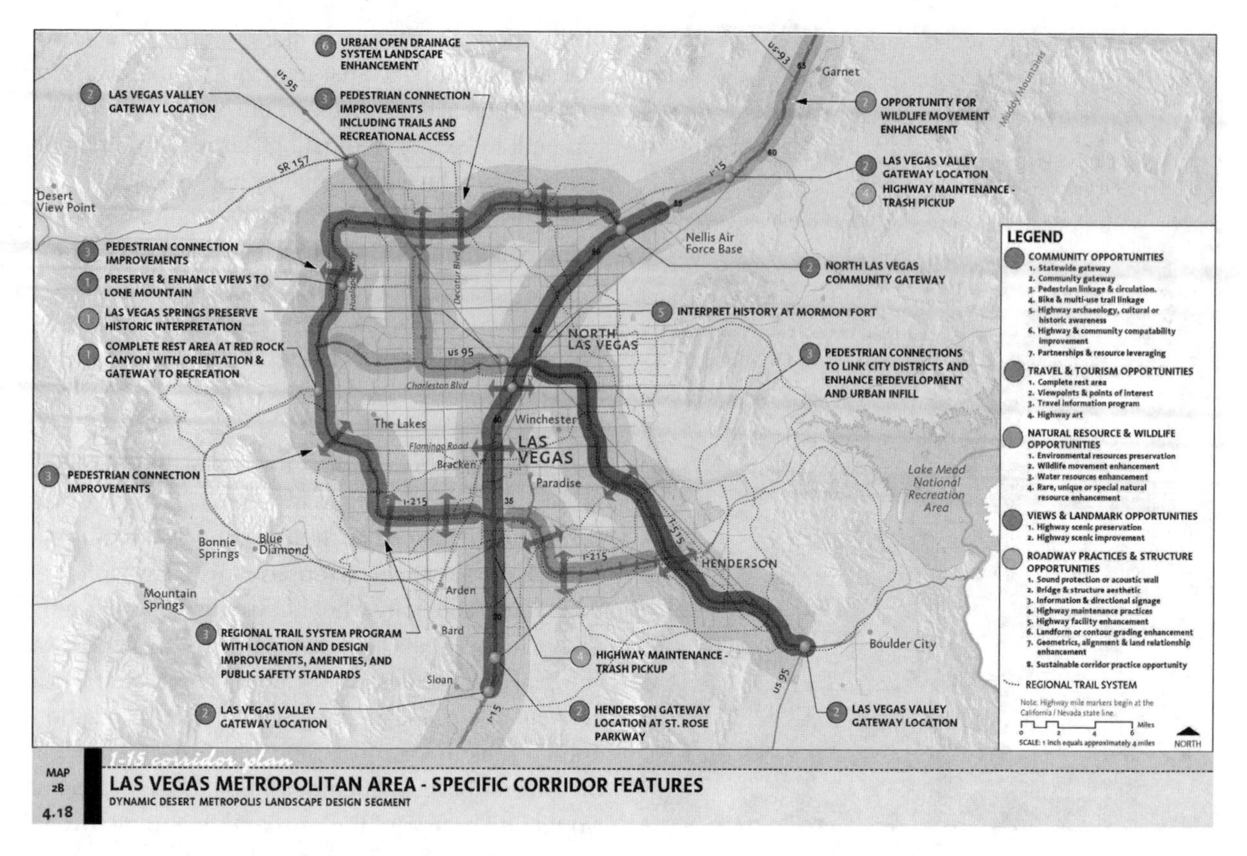

FIGURE 8-27 The Nevada DOT's I-15 Corridor Plan was developed to improve the visual appearance of I-15 with detailed illustrations synthesizing historic, current, and future conditions. Image courtesy of Design Workshop.

long to realize that we needed policies, procedures, and funding," recalls Mark Hoversten, former director of landscape architecture at the university.

The Landscape and Aesthetics Master Plan outlined a policy of integrating aesthetics into the design of all of the major highway projects in the state and provided a blueprint and a framework for Nevada DOT and the citizens of Nevada to turn their vision into a reality. The plan was completed in 2002, and the Nevada State Transportation Board adopted it as policy and authorized funds.

Corridor Plans

After the Landscape and Aesthetics Master Plan was adopted as policy, Nevada DOT began the process of implementing the plan. A design team led by the landscape architecture firm Design Workshop was hired to develop corridor plans for all of the major highways in the state. The first three corridors addressed were (1) I-15 corridor, including Las Vegas, (2) I-80 urban corridor, including Reno and Sparks, and (3) I-80 rural corridor east of Fernley. These are the three most visible and highly traveled corridors in Nevada.

The initial planning for each corridor focuses on producing an inventory of data, including history, settlement patterns, anticipated urban changes, travel and tourism, natural resources, wildlife habitat, viewsheds, landscape character, and applicable Nevada DOT standards and practices. The corridor plans define landscape types and a hierarchy of treatment

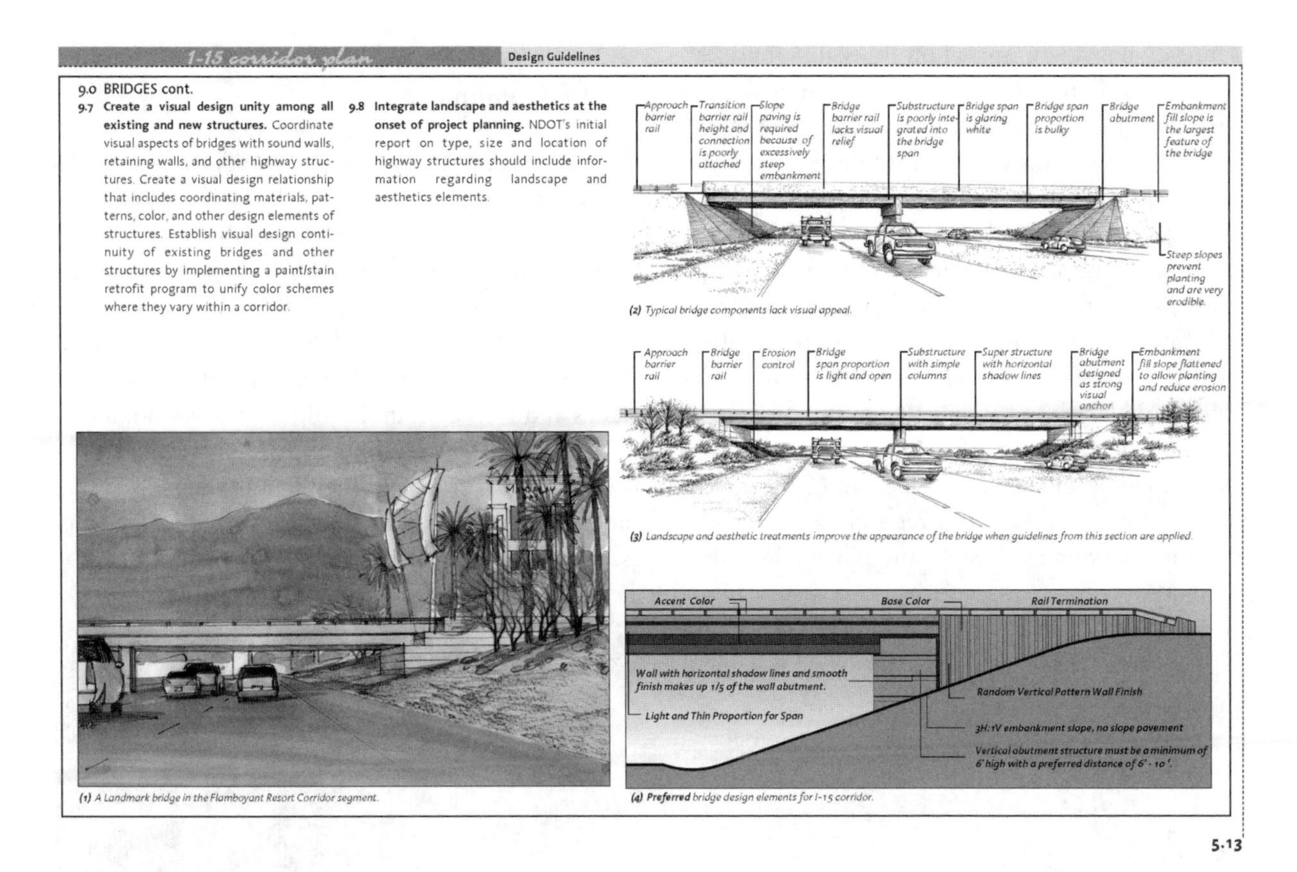
I-15 corridor plan — Design Guidelines

9.0 BRIDGES cont.

9.7 **Create a visual design unity among all existing and new structures.** Coordinate visual aspects of bridges with sound walls, retaining walls, and other highway structures. Create a visual design relationship that includes coordinating materials, patterns, color, and other design elements of structures. Establish visual design continuity of existing bridges and other structures by implementing a paint/stain retrofit program to unify color schemes where they vary within a corridor.

9.8 **Integrate landscape and aesthetics at the onset of project planning.** NDOT's initial report on type, size and location of highway structures should include information regarding landscape and aesthetics elements.

(2) Typical bridge components lack visual appeal.

(3) Landscape and aesthetic treatments improve the appearance of the bridge when guidelines from this section are applied.

(1) A Landmark bridge in the Flamboyant Resort Corridor segment.

*(4) **Preferred** bridge design elements for I-15 corridor.*

5.13

Figure 8-28 Included in the I-15 Corridor Plan is attention to the color and proportions of structures to improve aesthetic quality. Image courtesy of Design Workshop.

levels that Nevada DOT can apply to landscape segments with common characteristics. The treatments are arranged in a matrix from standard approaches to landmark approaches for the most striking and memorable landscape segments. Each level consists of combinations of treatments for softscape features, such as trees, shrubs, perennials, grasses, and ground treatments, and for hardscape features, including bridges, retaining walls, acoustic walls, pedestrian crossings, railings, barrier railings, lighting, and transportation art.

Funding

One of the keys to making the master plan successful is adequate funding. State policy now requires that up to 3 percent of the state's construction budget for new and capacity improvement projects be used to implement landscape and aesthetic treatments. The state also set aside $2 million annually for the retrofit of landscape and aesthetic improvements to rural highways, and local communities have contributed matching funds. The state is finding this approach a cost-effective way to ensure that new transportation projects avoid the problems associated with the Carson City Bypass.

Implementation

Nevada's Landscape and Aesthetics Master Plan has been implemented since the mid-2000s, and the plan is already having an impact on the state's transportation system. Two of the first projects include the Blue Diamond interchange and the Henderson Spaghetti Bowl. Over the course of the next decade or so, millions of dollars will be funneled through the program and spent on projects that improve the overall visual character of the state. "We have an incredible opportunity in this state, and everything we are doing today will have an effect on how highways are developed for the next 50 years," said Ron Blakemore, former supervising landscape architect with the Nevada Department of Transportation. "This plan is enabling [Nevada] to develop better projects and highways that not only look good but are safer and have a better fit with the environment" (Sipes and Blakemore, 2007).

Blue Ridge Parkway, Virginia and North Carolina

The Blue Ridge Parkway is a National Parkway, a National Scenic Byway, and an All-American Road that is considered by many to be the most scenic drive in the country. No book about green roadways would be complete without at least a short discussion of the Blue Ridge Parkway. The 469-mile corridor connects Shenandoah and Great Smoky Mountains National Parks and is located in the Blue Ridge Mountains of Virginia and North Carolina. The roadway meanders through the landscape, highlighting dynamic vistas and close-up views of the mountains and scenic landscapes of the Appalachian highlands.

FIGURE 8-29 The Blue Ridge Parkway offers some of the most spectacular scenery in the world. The area is home to more than a hundred tree species, a wide variety of wildflowers and shrubs, and fifty-nine species of birds. Image courtesy of the Library of Congress-HABS/HAER.

FIGURE 8-30 The spectacular fall foliage of the Blue Ridge Parkway presents a stunning array of brilliant yellows, oranges, golds, and reds, thanks to the great number of tree species. Image courtesy of the Library of Congress-HABS/HAER.

The parkway is the most visited unit in the national park system. In 2002, the parkway had more than 21.6 million recreation visits.

Work on the parkway began in 1935 near Cumberland Knob in North Carolina; construction of the entire project took more than fifty-two years to complete. Most construction was carried out by private contractors under federal contracts. State and federal highway departments, the Works Progress Administration, Civilian Conservation Corps workers, and immigrant stonemasons were also involved in building the parkway.

The parkway was designed for the recreational drive, and the speed limit is a little lower, the climbs a bit steeper, and curves a little tighter than on some other roads. In particular, the parkway uses descending radius curves that may get tighter as you go through it. This approach enables the road to fit the landscape and minimize impacts on environmental and cultural resources. The parkway uses short side roads to connect to other highways, so there are no direct interchanges with interstate highways or other major roads.

There are a number of visitor centers, campgrounds, scenic overlooks, picnic areas, trails and trailheads, and other recreation opportunities along the parkway. Many of these are closed in the winter, although the Folk Art Center in Asheville, the Blue Ridge Parkway Visitor Center adjacent to park headquarters, and the Museum of North Carolina

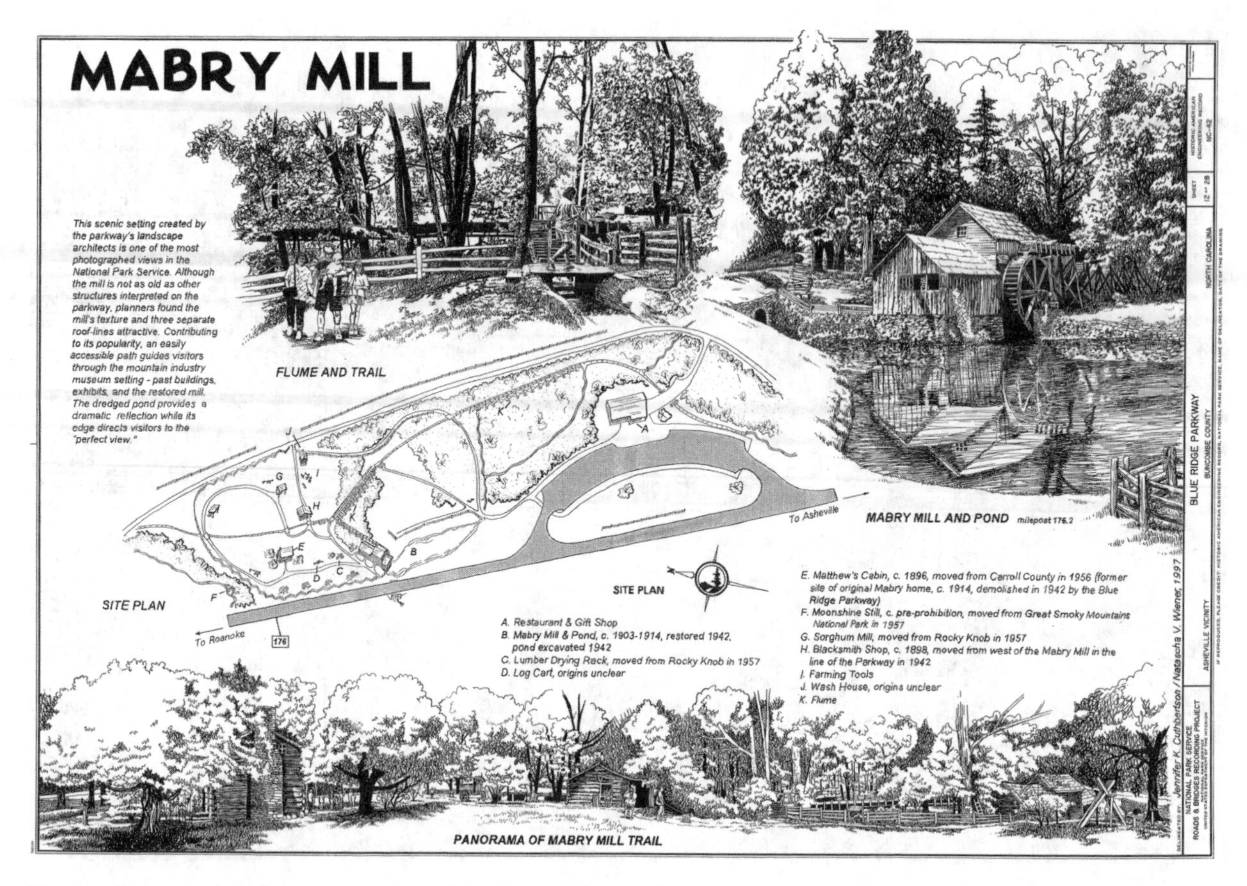

FIGURE 8-31 During the construction of the Blue Ridge Parkway, great care was taken while building the roadway to ensure that it blended into its natural surroundings. Image courtesy of the Library of Congress-HABS/HAER.

Minerals at Spruce Pine are open year round. The drive along the Blue Ridge Parkway is spectacular during the change in fall foliage, which begins around the middle of October.

Over time, the overall character of the parkway has changed. Since 1948, 75 percent of farmlands along the parkway have been converted to other land uses. As a result, there are more modifications along the corridor as a result of human activities.

Natchez Trace Parkway Bridge, Tennessee

Natchez Trace Parkway is a 450-mile road that connects Natchez, Mississippi, to Nashville, Tennessee. The Natchez Trace Parkway passes through the rolling forests of western Tennessee and then traverses the valleys of central Mississippi. The parkway was designed to follow the contours of the land, rising and falling with the surrounding hills and valleys.

One of the unique parts of the parkway is the Natchez Trace Parkway Bridge, which was built just outside of Franklin, Tennessee, in 1994. It is the first segmented concrete arch bridge built in the world. The beautiful bridge accents the surrounding Tennessee landscape beautifully.

The Natchez Trace Parkway Bridge is no ordinary bridge; it is a one-of-a-kind concrete arch bridge that uses two arches to support the bridge. This allows the bridge to be constructed using fewer vertical supports

FIGURE 8-32 The Linn Cove Viaduct snakes around the slopes of Grandfather Mountain along the Blue Ridge Parkway. Image courtesy of Wikipedia Commons.

FIGURE 8-33 With hundreds of parking areas and overlooks, Blue Ridge Parkway was designed so that motorists could enjoy a leisurely ride through the mountains. Image courtesy of the National Park Service.

FIGURE 8-34 The Blue Ridge Parkway tunnels, a total of twenty-six, were designed to have a minimal impact on the land. Image courtesy of the Library of Congress-HABS/HAER.

than most other bridges, and the supports are placed far apart and are not evenly spaced. This helps define the bridge as a visual landmark while also minimizing the visual and physical impacts on the landscape.

The first span of the bridge is 582 feet long and 145 feet tall, and the second is a bit smaller with a span of 462 feet and a height of 102 feet. The arches were precast in pieces near Franklin and brought to the site. Putting the bridge together took some innovative engineering. The bridge was built as a cantilever, held in place with steel cables, and built progressively, first one arch and then the other. The vertical supports were cast on-site concrete. The arches create a beautiful design for the bridge, and the smooth white concrete adds a stark contrast to the natural environment.

Creative Corridors, Winston-Salem, North Carolina

The Winston-Salem Creative Corridors project was inspired by a simple question: "Why can't the design of the highways that traverse our city create an image that is memorable?" The Creative Corridors project began in response to the North Carolina DOT's plan to create improvements on specific roadways in Winston-Salem.

Figure 8-35 The Natchez Trace Parkway is a drive through American history and natural beauty that extends from Mississippi to Tennessee. Image courtesy of Wikipedia Commons.

Figure 8-36 The Natchez Trace Parkway Bridge, located in Williamson County, Tennessee, is a 1,572-foot-long concrete double arch bridge spanning State Route 96 and a heavily wooded valley. Image courtesy of the National Park Service.

FIGURE 8-37 The Creative Corridors plan was created to rebuild a one-mile corridor of downtown Winston-Salem to reinvigorate the downtown city center. Image courtesy of the Design Workshop.

The Creative Corridors plan captures the process and outcomes of the master-planning effort by the design team led by Design Workshop from January through August 2011. The objective of this phase was to develop a plan guided by principles of aesthetics, urban design, place making, and sustainability.

The basic idea behind the Creative Corridors project is to develop each corridor so that it creates a unique setting for downtown, facilitates way-finding, and promotes transportation options such as pedestrian, bicycle, and mass transit. Specific themes have been identified and are how the unique brand

FIGURE 8-38 The guiding principles behind the Creative Corridors project are to create a network to unify the city. The project includes constructing new on-ramps on Business Route 40 and replacing eleven bridges. Image courtesy of the Design Workshop.

for Winston-Salem will be established: green, artful, iconic, and network.

The corridors can provide a series of visual cues that will communicate where the main gateways to the city are, highlight important landmarks and areas, and aid in navigation. They will also function almost as sculpture gardens, creating a calming, humanizing setting for the display of art and sound design principles, while also carrying out their main transportation function.

The Visionary Master Plan and Design Guidelines illustrate a detailed vision for improving the visual appearance of the built environment by the inclusion of public art as well as landscape, architectural, and engineering solutions that seek a high level of aesthetic integrity. The document provides a visual and textual story of the design analysis, definition, and discoveries that led to planning solutions and conclusions. It is intended for client use in presenting the project's vision to municipal and state officials for approvals, to attract the interest of the community and stakeholders, and to serve as the foundation for the next phases of the design process from which the plan will evolve.

The purpose of these design guidelines is to establish the acceptable aesthetic design vocabulary for the various elements that make up the corridors. As the design process proceeds beyond the master-plan stage, and after the final transportation plan and North Carolina DOT design requirements are finalized, these guidelines will continue to provide aesthetic direction and will help narrow the set of choices for subsequent design efforts.

Chapter 9

Natural Resources/Environmental Sustainability

A green roadway should be designed, built, and maintained so that it protects the natural environment, is respectful of people and place, and is part of an effort to achieve a sustainable future.

In combination with safety, the goal of green roadways is to protect existing natural resources. This means avoiding these resources by carefully considering where and how a roadway is developed. The horizontal alignment of a road can be shifted to avoid natural resources, and vertical alignments can be adjusted to minimize cut and fill, thus reducing the extent of impact. Retaining walls, geotextiles, and other methods can be used to protect existing natural resources.

At times, impacts to natural resources are unavoidable. When impacts do occur, emphasis should be on restoring or repairing the resources that have been damaged. The focus should be on maintaining a healthy, sustainable environment where natural processes continue to function as normal. There may even be opportunities to enhance existing conditions and improve environmental quality.

In recent years, the concept of environmental stewardship has increasingly gained acceptance. Environmental stewardship is the practice of not only protecting but enhancing the environment as a routine part of project development.

As stated in chapter 2, NEPA requires that natural resources be considered as part of any transportation project, and for good reason. Research has shown that a road's environmental footprint can be as much as twenty times the actual width of the right-of-way.

The level of environmental review varies widely, depending on the scale and impact of the project. Environmental permits have to be secured before a transportation project can be developed, and each permit process is unique and involves interagency coordination, information submission, possibly special public hearings, and specific forms or applications.

Wildlife Habitat

A green roadway respects wildlife. More than one million vertebrates, amphibians, reptiles, birds, and mammals are killed on roads each day in this country. For motorists, this is a safety hazard. For animals, it means disrupted migration and feeding patterns, destroyed or degraded habitat, and fragmentation. Instead of accepting this as the cost of doing business, we need to develop roads that reduce vehicle-wildlife collisions, preserve habitat, and maintain safe wildlife movement corridors.

Preservation starts with minimizing the extent of impact and protecting existing habitat areas before, during, and after construction. The many effects of roadways that can be detrimental to wildlife include habitat loss, habitat fragmentation, altered habitat quality, population fragmentation, and disruption of environmental processes.

Fragmentation is the subdivision of once large and continuous tracts of habitat into smaller patches. It can lead to several ecological processes that adversely affect wildlife populations, including edge effects and barrier effects. Edge effects include a range of detrimental ecological consequences that are associated with a decline in habitat quality, including invasive species, predators, and parasites. Barrier effects result in a physical or psychological restriction on movements of animals, and this eventually affects the health of a species. In particular, small animals such as frogs, turtles, and mice have a difficult time crossing roads safely and are vulnerable to barrier effects.

Roads can significantly degrade stream ecosystems by introducing high volumes of sediment into streams, changing natural stream flow patterns, and altering stream channel morphology. Changes in stream habitat affect the health of aquatic organisms. The survival rates of many salmonid species, for instance, decrease as fine sediment levels increase.

One approach that offers promise in helping ensure that roads take into consideration the environment is the emerging science of road ecology, which focuses on the corridor adjacent to the roadway and includes the land up to a mile or so on either side of the roadway. The result is a linear swath of land that includes the roadway as well as areas affected by the road.

Road Ecology

The science of road ecology is defined by Harvard professor of landscape architecture Richard T. T. Forman, lead author of the book *Road Ecology*, as the study of the interaction of organisms and their environment linked to roads and vehicles. Forman identified the following ecological concepts for corridors:

- The corridor should be as wide as possible. The corridor width may vary with habitat type or target species, but a rule of thumb is wider and larger areal extent is better.
- The longer the corridor, the wider it may have to be.
- It is preferable to have low intensity land uses adjacent to the corridor to minimize human impacts.
- To lessen the impact of roads, maintain as much natural open space as possible next to any culverts and bridge under- or overpasses to encourage their use by animals.
- Housing or other impacts should not project into the corridor, form impediments to movement, or produce harmful edge effects.
- If buildings or housing are to be permitted next to the corridor, establish a buffer and place a conservation easement over this area.
- Where the hydrology supports it, place the development's stormwater retention/detention facilities between the developed land and conservation land as an added buffer.
- Strict lighting restrictions should be developed for the houses adjacent to the corridor to prevent light pollution into the corridor.

As the understanding of road ecology grows, a multidimensional view of roads emerges, which takes into consideration not only the safety and efficiency of roadway networks to humans, but the ability to maintain ecological processes as well. This understanding requires participation and input from biologists, hydrologists, transportation planners, landscape architects, environmental planners, and engineers.

Wildlife Crossings

The basic idea behind wildlife crossing structures is to maintain landscape connectivity rather than try to redirect wildlife movements. There is evidence that appropriately sized and located wildlife crossings will extend the habitat of animals and decrease road kills. According to research by the Transportation Research Board, providing animal crossing structures and underpasses can reduce vehicle-wildlife collisions as much as 97 percent. (Morrall et al., 2007).

In most cases, wildlife crossings are installed on a project-by-project basis, and are designed for specific habitats, topography, and species. Most wildlife crossings are created in conjunction with highway upgrades, but they are constructed as part of new road projects as well.

Determining the best location will enhance the possibility that a crossing will be used. After identifying the animal populations affected by a road, locations for crossings can be determined by several methods, including the identification of good habitat and current wildlife movement patterns.

Animals often follow along water courses as a necessity for daily and life-cycle needs, so many crossings follow rivers and creeks. Water courses serve both upland and aquatic species. According to the U.S. Department of Agriculture, Natural Resources Conservation Service, more than 70 percent of all terrestrial wildlife species use riparian corridors.

Here are some examples of efforts to implement wildlife crossings:

- In 1973 Colorado installed the first underpass in North America for wildlife under I-70 near Vail Pass.
- The first set of multispecies crossings was built in the 1980s by Florida Department of Transportation (DOT) to accommodate the Florida panther and other species.
- Along the section of the Trans-Canada Highway that traverses Banff National Park in Alberta, Canada, wildlife have used the wildlife crossings more than 100,000 times in the past two decades.

Figure 9-1 Wildlife crossings, such as these underpasses in Utah, have proven successful in increasing driver safety as well as animal survival. Image courtesy of Utah Department of Transportation.

- Thousands of reptiles and amphibians were being killed annually where US Highway 441 crosses Paynes Prairie State Preserve in Florida until a three-foot-high wildlife barrier wall and culvert underpass system was constructed in 1999.
- In the I-90 Snoqualmie Pass corridor in Washington State, partners determined that "regardless of the build alternative," the project must connect habitat across I-90 for fish and wildlife.
- The Linking Colorado's Landscapes project seeks to identify and prioritize wildlife habitats and corridors across the state. The first step was to analyze the entire state, and this process resulted in the identification of thirteen key wildlife-crossing areas.
- The Wyoming Wildlife Action Plan modeled the condition of wildlife habitats across the state and identified fifty-two terrestrial ecological systems across seven ecoregions, aggregated into seven major community types.

FIGURE 9-2 This proposed land bridge across I-90 in Washington State would enable wildlife to cross over the road. This type of approach helps link the forested habitats that are separated by the highway. Image courtesy of Washington Department of Transportation.

Crossings include culverts, bridge structures, and over-crossings that allow wildlife safe passage under or over the roadbed. As a general recommendation, crossings should be sized to accommodate the largest and most wary species in the vicinity.

Overpasses include (1) landscape bridges designed exclusively for wildlife use; (2) smaller wildlife overpasses; (3) multiuse overpasses that allow for both wildlife and human use; and (4) canopy crossings that are designed specifically for semi-arboreal and arboreal species that use canopies for travel.

Underpasses include (1) viaducts with large span and vertical clearance; (2) large mammal underpasses designed specifically for wildlife use; (3) multiuse underpasses that are used by both wildlife and humans; (4) underpasses with water flow that accommodate aquatic species as well as terrestrial wildlife; (5) small to medium underpasses designed for smaller mammals; (6) modified culverts with dry platforms or walkways for smaller mammals; and (7) amphibian and reptile tunnels.

For US Highway 93, which is located on the Flathead Indian Reservation in western Montana, more than fifty wildlife crossings were designed for a forty-two-mile stretch of the reconstructed highway. Many of the crossing structures are large (twelve feet by twenty-two feet) culverts, but there are also several bridge structures that have spans ranging from eighty to four hundred feet. Bridges were located to allow for the easiest wildlife movement, spans are wide enough to allow dry land passage on one or both sides of creeks, and clearance under the structures was set to allow elk to move underneath with ease. There is also a wildlife overpass, where the road is located in a tunnel and wildlife are allowed to move across a "land bridge." For more on this project, see the case study on page 195.

FIGURE 9-3 The wildlife underpass at Gold Creek in Washington State not only benefits the local terrestrial wildlife, but also benefits the animals living in the creek. Image courtesy of Washington State Department of Transportation.

FIGURE 9-4 Cameras show how these river otters are using a tunnel to safely cross a busy roadway. Image courtesy of Montana Department of Transportation.

FIGURE 9-5 Deer can safely cross US Highway 93 that might otherwise interfere with their normal migration route. It is important to make the tunnels as wide as possible with a dirt and rock floor. Image courtesy of Montana Department of Transportation.

Habitat Alterations

The basic idea behind habitat alteration is to attract or repel wildlife from transportation corridors. In other words, we want wildlife to go where it is safe, and we want to prevent conflicts with motorized vehicles.

One concern is that by planting or preserving wildflowers, shrubs, groundcovers, and native grasses near roadways, we may be encouraging more wildlife to use these areas. Plantings around the travel lanes should be selected to ensure that we don't inadvertently attract species closer to roadways.

Mowing should be minimal in environmentally sensitive areas to preserve habitat of ground-nesting birds. Many states use a three-tiered approach to managing roadside vegetation. The areas adjacent to the roadway are divided into three different zones, with each being mowed to different standards. Reducing the frequency of mowing lowers cost and increases plant species diversity.

In Wisconsin, DOT workers are careful to avoid mowing wild lupine plants within the right-of-way because the plant is a primary source of nourishment for the Karner blue butterfly, which is a state-endangered species. Iowa's Living Roadways program emphasizes environmentally friendly techniques to manage roadside areas. In Iowa, native grasses, trees, and shrubs are planted and allowed to flourish along roads and trails.

Figure 9-6 Natural vegetation near the opening of a wildlife crossing will give animals the security of their preferred environment and can work as a guide to motivate them to use the crossing. Image courtesy of Jones & Jones.

The result has been the restoration of Iowa's native prairie habitat, a reduction in soil erosion and sedimentation, and beautification of the Iowa landscape.

Planning for Climate Change

In 2008, the Transportation Research Board published *Potential Impacts of Climate Change on U.S. Transportation*, which states that the solution for dealing with climate change will be a combination of two strategies: adaptation and mitigation. The adaptation strategy means adjusting to the climate changes that will occur. The mitigation strategy means reducing the severity of climate change by reducing greenhouse gas emissions (Snelling, 2010).

Traditional transportation planning processes are inadequate for dealing with the upcoming changes associated with climate change. Innovative planning approaches that promote sustainability and flexibility could significantly reduce the severity of impacts associated with climate change. In particular, transportation policies need to reduce energy consumption, travel demand, and dependence on oil.

Carbon considerations will fundamentally change how we look at transportation projects. At the state level, mandates are being put in place to establish rules and regulations for dealing with climate change. For example, California's Sustainable Communities and Climate Protection Act (SB 375) is the nation's first legislation to link transportation and land use planning with climate change. In other states, stricter building regulations and a push for carbon neutrality will drive demand for more sustainable transportation planning.

Open Space Opportunities

The U.S. DOT estimates that we are losing approximately two million acres of land to development each year as a result of transportation projects. In addition, a study in the journal *PLoS ONE* found that almost 3.5 million acres of open space were lost to urban sprawl in the United States from 1990 to 2000 (McDonald et al., 2010). If these trends continue, many of the nation's treasured natural areas and open spaces will disappear as vast tracts of land are developed into urban areas in the next couple of decades.

Parks, greenways, open space, and recreation services are integral parts of the fabric of a healthy community. Each serves to strengthen structure, culture, connectivity, and unique sense of place for residents and visitors. In most surveys about quality of life two elements are almost always at the top of the list of concerns: transportation and open space.

One major reason for conserving open space is to preserve natural resources and maintain the functionality of environmental systems that provide benefits such as food and clean water and air. The cost to replicate these systems is unimaginable; to create the flood control provided by natural wetlands, for example, would cost more than a hundred times that of protecting the naturally occurring resource (Hitchcox, 2001).

Since 1992, Transportation Enhancement activities have contributed $5.6 billion in federal funds to support more than fourteen thousand projects for paths, trails, and bicycle facilities. The Recreational Trails Program provides money to states to develop and maintain recreational trails and trail-related facilities.

The North Moab Recreation Area is located in the City of Moab, Grand County, Utah. Tourism is Grand County's most important economic resource. Many of Moab's 2.5 million visitors come because of ample opportunities to bike and hike. To help alleviate problems with too many cars, the North Moab Recreation Areas Alternative Transportation Project was initiated. It is an integrated motorized and nonmotorized transit system that includes two transit hubs served by private shuttle businesses, 42.5 miles of bike paths and lanes, and a bicycle/pedestrian bridge across the Colorado River.

The Paul S. Sarbanes Transit in Parks Program provides funding for alternative transportation systems, such as shuttle buses, rail connections, and even bicycle trails, for national parks and other federal lands. The purpose of the program is to conserve natural, historical, and cultural resources by reducing the impacts of motorized vehicles. The program is administered by the U.S. DOT, the Department of the Interior, and the U.S. Forest Service.

Protecting Open Space

One of the best ways to protect open space is to minimize the construction of new roads in "greenfields," which are undeveloped areas. Reducing travel demand, providing alternative modes of transportation, adding travel lanes within existing rights-of-way, and emphasizing smart growth options help reduce the need for new roads in greenfields.

If new roads need to be developed in greenfields, reducing the footprint of the road will also reduce potential impacts to existing cultural and natural resources. This can be done by having fewer travel lanes, narrower shoulders, and less cut and fill.

Another approach is to reclaim open space by decommissioning existing roads. Cities around the world are replacing urban highways with surface

Figure 9-7 Moab, Utah, is surrounded by stunning red rock landscapes, beautiful scenery, and perfect climate, making the area ideal for year-round outdoor events. Image courtesy of Wikipedia Commons.

Figure 9-8 The Brandywine Valley Scenic Byway is a 12.2-mile unforgettable journey through historic sites, magnificent gardens, museums, and estates in Delaware. Photo by Rick Darke, courtesy of National Scenic Byways Online (www.byways.org).

streets, saving billions of dollars on transportation infrastructure and revitalizing adjacent land with walkable, compact development and creating new open space.

Water Resources

Water is our most important environmental concern. There are a number of water-related issues associated with transportation projects, including stormwater runoff and water quality. More paved surfaces lead to increased stormwater runoff, which can result in flooding, erosion, and environmental impacts. Both the construction and subsequent use of roadways and associated development can have a negative impact on water quality as a result of sedimentation and increased pollutants.

A fundamental problem with traditional approaches to addressing stormwater is that this water has been treated as waste, and the idea was to collect the water and get rid of it as quickly as possible. We also treated water as if we had a right to use it and as if we would never run out. As a result, many rivers, streams, lakes, and underground aquifers have been seriously affected by human activities.

The U.S. Environmental Protection Agency (EPA) considers urban runoff and pollution from other diffuse sources the greatest contaminant threat to the nation's waters, and much of this runoff is due to our transportation infrastructure. More than 235,000 river miles in the United States have been channelized, 25,000 river miles have been dredged, and another 600,000 river miles are impounded behind dams. Nearly 40 percent of the rivers and streams in the United States are too polluted for fishing and swimming.

The primary objective for stormwater management is to follow natural processes as closely as possible. This type of green approach is not only the most environmentally friendly, it is also the most effective in terms of functionality. There are two basic approaches to stormwater management: conveyance and infiltration. The conveyance approach focuses on getting rid of the water as quickly as possible. Virtually every city in the United States has a conveyance approach that uses underground pipes, drainage ditches, and curbs and gutters. The benefit of using a conveyance approach is that it reduces immediate flooding problems. The downside is that by diverting all the water, we create problems downstream. The increased volume of water can lead to severe erosion problems, downstream flooding, and sewage overflow.

Total maximum daily load (TMDL) is the maximum amount of any pollutant, contaminant, or impairment that can enter a body of water before the

FIGURE 9-9 The "no dumping" sign by a drain helps to prevent water pollution. Image courtesy of EDAW.

quality of the water is deemed unfit for its designated uses. TMDLs are emerging as a primary issue for DOT programs as more water bodies are listed as "impaired." Stormwater runoff can cause serious problems with TMDL for transportation projects.

Wetlands

Wetlands provide a number of benefits, including improving water quality; providing animal habitat, sediment filtration, and temporary water storage; reducing pollution and potential flood damage, producing oxygen, and affecting nutrient recycling. Wetlands are also among the most productive ecosystems.

Historically, wetlands were considered wastelands with little if any economic value. EPA estimates that more than one hundred million acres of wetlands have been destructed in the United States since the late 1700s. Fortunately, the loss of wetlands has gradually been reduced, and the U.S. Fish and Wildlife Service indicates that wetland gains exceeded losses between 1998 and 2012.

Wetlands and buffers of natural vegetation adjacent to water bodies can retain floodwaters and remove contaminants contained in surface runoff.

FIGURE 9-10 The Bossu Wetland is a sixteen-acre farmland in Ontario that has been restored to a wetland, creating a safe home for wildlife that would have otherwise not been able to survive. Image courtesy of EDAW.

Floodplains along rivers, streams, or creeks are typically inundated with water following storms. Floodplains help reduce the number and severity of floods, filter stormwater, and minimize nonpoint source pollution. Water expands into the floodplain areas and infiltrates into the ground, slowing water flow and allowing groundwater recharge.

Riparian corridors include grass, trees, shrubs, and a combination of natural features along the banks of rivers and streams. Protecting these corridors is critical for protecting water quality.

National Pollutant Discharge Elimination System–Much of transportation-related water quality work is structured to comply with the National Pollutant Discharge Elimination System (NPDES) permitting program established under Section 402 of the Clean Water Act. NPDES permitting functions as the primary regulatory tool for ensuring that state water quality standards are met. NPDES is not directly related to creating green roadways, but because roads have to meet water quality standards, design decisions that address water often result in a more environmentally friendly roadway. NPDES permits, issued by U.S. EPA or an authorized state agency, contain discharge limits intended to meet water quality standards and national technology-based effluent regulations (Sipes, 2010).

NPDES permits regulate the discharge of pollutants from point sources, such as pipes, outfalls, and conveyance channels. In general, facilities that discharge wastewater into water bodies are required to have a permit under the NPDES program. To meet NPDES requirements, each local stormwater program is responsible for establishing a stormwater management plan, which gives specific local requirements targeted to meet the environmental needs of each watershed.

Water Solutions for Green Roadways

A green roadway should seamlessly integrate permanent stormwater practices into the roadway environment in an attempt to replicate natural hydrologic processes. Every piece of the roadway should be designed to serve a water quality purpose. As stormwater is directed off the roadway and its collection systems, innovative best management practices can be used to maximize treatment efficiency and protect the natural environment. Best management practices include a variety of methods that have been effective in past applications.

The design of stormwater management facilities must be handled very carefully along roads that are surrounded by or adjacent to cultural, natural, or visual resources. This is particularly true of historic roads with tight rights-of-way, narrow shoulders, and trees or walls that are close to the road.

Possible solutions for reducing stormwater runoff and improving water quality include the following:

Basins and Swales

Rainwater harvesting–Rain gardens are small bioretention ponds or depressions that slow down stormwater and allow it to percolate into the soil. The basic idea of a rain garden is to capture stormwater runoff from impervious areas such as roofs, driveways, walkways, parking lots, and compacted lawn areas and divert it into vegetated areas instead of having it run off into the storm sewer system. A big advantage is that less water flows into storm sewers, so less piping and infrastructure are needed.

Stormwater retention/detention best management practices–Retention or detention best management practices involve the control of stormwater by gathering runoff in wet ponds, dry basins, or multichamber catch basins and slowly releasing it to receiving waters or drainage systems.

Sediment controls–Sediment basins capture sediment that can be transported in runoff. Filtration and gravitational settling during detention are the main processes used to remove sediment from urban

FIGURE 9-11 This bioswale in Portland, Oregon, is designed to remove silt and pollution from surface and runoff water. Image courtesy of Kevin Robert Perry, City of Portland.

runoff. Sediment basins slow down stormwater runoff and allow sediment to settle out of the water.

Bioswales–Wet and dry swales are constructed to handle stormwater runoff on the surface. A grass or vegetated swale slows down water and reduces pollution carried by stormwater. These are similar to rain gardens in that they are contoured vegetated areas intended to slow and filter stormwater. Technically they differ in that they typically convey a positive flow of water during storm events without pooling or ponding in depressions.

Biofiltration systems–Stormwater that passes through a biofilter is collected in a structural underdrain and discharged through a polyvinyl chloride underdrain connected to a raised drop inlet.

Bioslopes–Bioslopes are a version of biofiltration. They contain the same elements as biofilters. Roadway runoff is intercepted and filtered through the high-flow media, which is a combination of soil and gravel aggregate, before being discharged through a polyvinyl chloride underdrain. Flows in excess of the design storm simply bypass the bioslope as sheet flow.

Bioretention basins–This practice involves retaining, infiltrating, and filtering stormwater runoff through a vegetated depression.

Stream buffers–Stream buffers are strips of riparian vegetation along waterways that physically separate and protect the water and associated wildlife from land development activities. These buffers prevent stormwater, which carries sediments and other harmful contaminants, from flowing directly into waterways. Buffers can also prevent unstable stream banks from falling in and eroding.

Naturalized channel design and infiltration methods–Where possible, avoid paving drainage ditches or check dams with asphalt or concrete. On a case-by-case-situation, use geotextiles, impervious mats,

Figure 9-12 A bioretention basin is a landscaped depression that directs stormwater into an area where it can be treated by a number of physical, chemical, and biological processes. Image courtesy of EDAW.

or stone lining to maintain a naturally appearing channel.

Low-impact development–This type of development relies on vegetation to control water flow and quantity.

Pavement for Stormwater Management

Pervious pavement–Pervious pavement allows post-construction stormwater infiltration rates to remain the same as before construction began. Stormwater also can be filtered through porous asphalt, so sediment and other pollutants are captured in the porous matrix, preventing their discharge from the roadway. (See "Pervious Paving" in chapter 10.)

Green shoulders–Green shoulders, which are reinforced grass surfaces, can reduce the velocity of stormwater runoff from a road.

Curb removal–Removing curbs from roadways allows runoff to filter through vegetated shoulders or infiltrate into the ground. Where curbs are necessary, curb breaks can be used.

Management Practices

Landscape irrigation–When landscape irrigation is needed, efforts could be made to reduce use of potable water. Water used could be restricted to captured rainwater, recycled wastewater, recycled gray water, air-conditioner condensate, blow-down water from boilers and cooling towers, or water treated and conveyed by a public agency specifically for nonpotable uses. Flow meters could be installed to record and monitor water use in the landscape irrigation areas.

Erosion and sediment control measures and practices–Temporary and permanent erosion and sediment control measures should be established and implemented at the earliest practicable time consistent with good construction and management practices.

Control of litter and debris on roadsides–Roadside litter control practices can improve runoff quality by limiting trash in runoff conveyance and treatment systems and receiving water bodies.

Manage pesticide use–Overapplication of pesticides may cause excess chemicals to leach to groundwaters or flow into surface waters.

Air Quality

Automobiles cause air pollution. Air pollution can damage ecosystems, reduce quality of life, contribute to climate change, and increase health problems for people and wildlife.

In 2011 cars and trucks contributed 450 million metric tons of carbon dioxide and 60 million metric tons of carbon monoxide per year, which is 62 percent and 32 percent of the U.S. totals. Cars and trucks also emit approximately half of all carcinogenic and toxic air pollutants. To make matters worse, the federal Department of Energy projects that carbon emissions in the United States will continue to increase by 1 percent per year, with transportation sources growing 20 percent faster than the average (NRDC, 2000).

According to the U.S. EPA, more than thirty-five million people in the United States live within three hundred feet of a major road. A major concern is the impact that air pollutants have upon the health of these people. Studies have shown that people who live, work, or spend much time near major roads have increased health problems related to air pollution from roadway traffic. These problems include reduced lung function, asthma, cardiovascular disease, low birth weight, preterm newborns, and premature death.

The Clean Air Act of 1970 is the primary law for protecting and improving the nation's air quality and the stratospheric ozone layer. Since then, additional laws and regulations have been added, including the 1990 Amendments to the Clean Air Act.

The amount of pollutants in the air is measured to determine if air quality meets defined standards. If standards are not met, fines or injunctions can be leveled against the responsible agencies.

Improving Air Quality

There are many ways that we can address air quality issues associated with roads. The basic hierarchy to control the dust, emissions, and air pollution is by (1) prevention, (2) suppression, and (3) containment.

As a general rule of thumb, every automobile trip that is eliminated from our roadway improves our overall air quality by eliminating vehicle emissions. For example, according to Ken Kifer's website (www.kenkifer.com), the average car emits about six tons of carbon dioxide every year. In comparison, an average cyclist might emit thirty to fifty pounds of carbon dioxide per year while cycling.

We have broken potential air quality improvements into four major categories: policies, materials, construction, and infrastructure.

Policies

Policies can be implemented to control what happens during the construction process and during the operation of a roadway. These policies are typically determined either at the state level, when the issues addressed are consistent across the state, or at a local level by a county, municipality, or organization. Examples of policies include:

Figure 9-13 Trees and grass shoulders help define the overall rural character of US Highway 231 in Virginia. Photograph by Cate Magennis Wyatt, courtesy of National Scenic Byways Online (www.byways.org).

- Require adjacent commercial and industrial users to provide a dustless surface on which vehicles drive.
- Establish practices for public agencies that will improve air quality. These could include (1) no vehicle idling, (2) electric or hybrid vehicles, where feasible, (3) bike patrols, (4) use of global positioning system technology to allow more efficient trips, (5) more business online, and (6) telecommuting opportunities for employees.
- Implement land use policies and promote mixed-use developments that create walkable communities and encourage people to park their cars. (See "Land Use Planning, Smart Growth, Complete Streets, and Transportation Infrastructure" in chapter 4.)
- Increase capital programs for public transit.
- Develop programs to limit or restrict vehicle use in downtown areas or other areas of emission concentration, particularly during periods of peak use.
- Encourage alternative fuels that are more efficient and have lower emissions.
- Define clean air zones where tighter air quality requirements are implemented.

Materials

- Reduce the impact of materials on air quality during the construction process. Encourage use of materials that actually improve air quality.
- Use new paving materials that can improve air quality by absorbing carbon dioxide. For example, porous concrete with olivine, a green-colored mineral, as a supplement material can absorb ten times more carbon dioxide from the

air than the amount of carbon dioxide generated in concrete production. Cement with magnesium silicate as an admixture can also absorb carbon dioxide as it cures.
- Incorporate native plantings along the road corridor. Plants are obvious buffers along roads because they use carbon dioxide as they grow.

Construction

- Take measures to reduce air pollution during construction.
- Implement dust control measures.
- Schedule and manage construction to reduce congestion and lane closures.

Infrastructure

- Provide facilities that offer alternative means of transportation. These include adding transit stops and encouraging carpools, park and rides, and other facilities that promote ride-share programs.
- Develop hike and bike trails that encourage people to walk or bike, both for recreation and for commuting to work.
- Provide roads or lanes solely for use by passenger buses or high-occupancy vehicles.
- Implement highway ramp metering, traffic signalization, nonstop tolls, efficient interchanges, and related programs that improve traffic flow and achieve a new emissions reduction.
- Provide fringe and transportation corridor parking facilities serving multiple-occupancy vehicle programs or transit operations.
- Establish programs and vehicle information systems for breakdown and accident scene management and nonrecurring congestion to reduce congestion and emissions.

Energy Conservation

One of the objectives of green roadways is to be as energy-efficient as possible. Transportation accounts for approximately 25 percent of world energy demand and for more than 62 percent of all the oil used each year. Road transportation alone is consuming on average 85 percent of the total energy used by the transport sector in developed countries.

Reducing Energy Usage

The relationship between transport and energy is a direct one. A significant amount of energy goes into producing materials for new roads and into constructing and maintaining these roads. Low-energy–consuming construction and maintenance techniques, such as prefabrication and low-temperature asphalt, can help reduce energy usage. (See "Green Construction Materials" in chapter 10 for additional information on energy alternatives.) Roadways can also be designed to be more efficient, so vehicles aren't consuming energy by sitting in congested traffic. The energy associated with producing automobiles and the fuel needed to power them is also an issue, but that is beyond the scope of this book. The resource list includes some sources on that topic.

There are a number of ways to reduce energy demands associated with road construction and maintenance:

Carbon-neutral solutions–Finland is looking to develop an eighty-one–mile stretch of road near the Russian border as a carbon-neutral green highway. The road would link the cities of Turku and Vaalimaa and would include charging stations and biofuel stations as well as traditional gasoline and diesel fueling stations. Initial planning has begun, and the project could be constructed as early as 2016 (Squatriglia, 2010).

Renewable energy installations–These installations, typically referred to as REIs, are commonly used in rural sites where utility services are not available, or in locations where there is an effort to reduce energy consumption. New Mexico, Florida, and other DOTs began to use photovoltaics to power roadside appurtenances in remote settings starting more than two decades ago.

Some solar-powered lights have the solar panel, battery, and light combined into a single fixture. Fixtures can include a motion sensor that communicates wirelessly with other nearby fixtures to synchronize low-high activation when one of the lights within the network senses motion. For example, the 3i crosswalk in-road lighting system reduces accidents by up to 80 percent because it alerts drivers that pedestrians need to cross the street. The system consists of solar-powered in-road light-emitting diode (LED) markers activated by in-pavement tiles that detect pedestrians.

Motion sensors–Although roadways are often busy, there are many times when few if any cars are present and electrical systems are left turned on unnecessarily. To minimize the electricity used by lights, signage, and other uses, a wireless vehicle detection system could be used to activate electrical systems by identifying the presence and movement of vehicles. Wireless vehicle detection systems utilizing magnetometer technology can be used to detect the presence and movement of vehicles.

LED lighting–Significant advances have also occurred in intelligent low-power lighting solutions. Primary among these are LED lights, which consume significantly less electricity than traditional metal halide or high-pressure sodium lights and provide superior color, uniformity, and lumen output. In addition, because LED elements have a rated life typically between eight and eleven years of continuous use, long-term maintenance costs are greatly reduced.

Carbon sequestration–The Federal Highway Administration (FHWA) is studying the use of vegetation within highway rights-of-way to reduce and sequester greenhouse gas emissions. The New Mexico DOT is undertaking a four-year, $2-million research project to quantify the amount of atmospheric carbon that grasslands along highway rights-of-way can sequester. Options for divestiture are (1) selling carbon credits on an appropriate greenhouse gas market or registry for revenue, (2) using carbon credits to offset the DOT's emissions, or (3) using the credits toward meeting statewide objectives.

Intelligent transportation systems–These systems are now being incorporated into many new green roadway projects. Technologies that are part of intelligent transportation systems programs include low-power dynamic messaging signs and road weather information systems. Solar-powered systems support several "actionable" weather sensors, including wind speed and direction, precipitation, visibility, optical road surface status, and water depth. For toll roads, intelligent transportation systems applications can be used to identify each vehicle, electronically collect the toll, and provide general vehicle/traffic monitoring and data collection and information for incident management and alternate route guidance.

Generating Energy

Transportation rights-of-way and roadways can be used to generate energy. The more than sixty million acres of land within highway and road rights-of-way could be used for bioharvesting, solar panels, and other similar uses.

Solar energy–Solar energy is becoming more affordable and advances in technology indicate that it will soon challenge more traditional sources of energy. One benefit is that 87 to 97 percent of the energy produced by the solar arrays will be free of pollution and greenhouse gas emissions.

Solar arrays within highway rights-of-way have been generating electricity in Europe for decades. In 2008, the first solar arrays along a U.S. highway were installed by the Oregon DOT. The project covers about eight thousand square feet and is roughly the length of two football fields. It cost $1.3 million to construct and produces 112,000 kilowatt hours of energy per year.

New Jersey's biggest utility is outfitting with solar panels 200,000 utility poles in and along road rights-of-way.

Solar panels at park and ride facilities could allow the installation of quick-charge stations for electric cars. These types of stations can charge a Nissan Leaf from empty to 80 percent charge in less than thirty minutes. California and Japan are already using these types of stations, and many urban areas within the United States are considering following suit. Caltrans plans to install $20 million in new solar energy systems at seventy of its facilities throughout the state.

Electric highway–Washington Governor Christine Gregoire has promoted the idea of the nation's first electric highway along I-5 in Washington. In June 2010, Gregoire said, "The transition to clean energy will have a big impact on Washington's future economic growth. The electric highway offers significant new business opportunities, attracting automakers and clean technology industries that are looking to locate or expand in an EV-ready [electric vehicle] state." Through the Electric Highway Project, the state will partner with private companies to install fast-charging infrastructure in critical charging zones in underserved locations along major interstates. This electric highway project supports the state's efforts on the West Coast Green Highway, a tristate initiative to promote the use of cleaner fuels along I-5 from British Columbia to Baja California.

Route 66, which connects Chicago to Los Angeles via St. Louis and Joplin, Missouri; Amarillo, Texas; Albuquerque, New Mexico; and Kingston, Arizona, is one of the most popular travel routes in the country for those interested in driving for pleasure. A combined effort by the Electric Highway, Route 66 Alliance, and the Green Roadway Project focuses on the installation of solar panels, geothermal devices, and other renewable-energy generators along Route 66 to promote the idea as part of an "electric highway."

Thermal energy–Thermal energy is a form of ambient energy available from natural heating of pavements. A type of Portland cement has been developed that can capture ambient thermal energy available from pavements.

Piezoelectric generators–A series of in-roadway piezoelectric generators can be used to harvest wasted mechanical energy imparted to the pavement by converting mechanical stress into current or voltage. For every one thousand feet of installed piezoelectric generators, approximately 150 kilowatt hours of electricity can be harvested every hour under normal to heavy traffic conditions. The construction costs and the expected return on investment time are estimated to be much lower than for solar energy. The technology is applicable to any place with heavy vehicle travel and is not confined to a specific climate or geographic area, as are solar and wind energy. Piezoelectric generators may be installed in new roadways or while resurfacing existing roadways. They have an expected service life of thirty years.

Kinetic energy–PaveGen is a recycled rubber paving slab that generates kinetic energy when people step on the paver. The first commercial application was in London, England, near the 2012 London Olympic stadium. The twenty tiles installed are expected to generate about half the electricity needed to power outdoor lighting for a nearby mall. The slabs compress five millimeters when stepped on, and this is enough to convert absorbed kinetic energy into electricity (Webster, 2011).

Generating biomass–Some DOTs are exploring biomass harvesting in the roadway right-of-way. Biomass crops are plant materials used to make electricity or fuels. Biomass crops are often divided by "generation." Research indicates that first-generation crops such as corn for ethanol or soybean or sunflower seeds for biodiesel probably cannot be grown in highway medians because they require intensive farming practices. Third-generation crops such as fast-growing salix willow hybrids, switchgrass, and miscanthus, on the other hand, could be grown in highway rights-of-way.

Wind power–In wider rights-of-way, it is possible to use wind turbines running parallel to a roadway to generate energy. Wind power is one of the most environmentally friendly methods for producing electricity, but it is highly visible. Wind turbines are now generally much quieter and more visually attractive, but there are still concerns about the visual impact they have upon the landscape. Town, regional, and state planners want to be able to protect the most scenically valuable or sensitive areas but also to encourage wind development. In states where tourism is a major part of the local economy, the visual quality of communities and adjacent landscapes is important.

In 2010, the National Renewable Energy Laboratory investigated potential areas for wind development in the continental United States, and identified sixty thousand prime sites. Some of these sites overlap with transportation corridors.

Case Studies—Integrating Natural Resources and Green Roads

US64/Corridor K, Cherokee National Forest, Tennessee

The US64/Corridor K project, in south central Tennessee, is in the study phase to find potential new routes on or near US Highway 64 through the Ocoee River gorge. The project, which would either widen the existing roadway or develop a new alignment, has been on the books for decades, but interest in the project increased after a 2009 rock slide closed the

FIGURE 9-14 Wind turbines are an alternate energy source that use kinetic power to generate electricity. Image acquired from depositphotos.com.

Figure 9-15 The purpose of the US64/Corridor K project is to improve US Highway 64/Corridor K through the Ocoee River gorge in Polk County, Tennessee. Image courtesy of J. Sipes.

highway for about eight months. The project is an example of the importance of considering environmental, cultural, and visual resources as early as possible in the planning process.

US64/Corridor K is part of the larger Appalachian Regional Commission's Appalachian Development Highway Program, which was authorized by Congress in 1965. This highway system stretches from New York to Mississippi, with each corridor being named from A to X. US64/Corridor K is one of the unfinished pieces of the highway.

US64/Corridor K, as proposed, would traverse the Cherokee National Forest and would be close to the Ocoee River, one of the most popular tourist areas in the state. Because a new highway would have a significant impact on the forest, the U.S. Forest Service played a major role during the National Environmental Policy Act (NEPA) process, including defining the purpose and need. In particular, the Forest Service wanted to emphasize the importance of environmental resources and ensure that visual resources were also considered.

The final purpose and need, as defined in the project's environmental impact statement, is as follows: "To implement a safe, reliable and efficient east-west transportation route from just west of where

Figure 9-16 A portion of US64/Corridor K is part of the Ocoee Scenic Byway, which was designated as the nation's first National Forest Scenic Byway. Image courtesy of J. Sipes.

US 64 crosses the Ocoee River, traversing Cherokee National Forest, to SR 68, that improves regional transportation linkages and protects/enhances environmental, cultural, aesthetic and historic resources as well as recreation opportunities. It is also to support objectives of local, regional, state and federal plans, land use and transportation goals, and economic development in the SE region of Tennessee."

In 2008, the Tennessee DOT, collaborated with local, state, and federal agencies and the public to conduct a corridor-level planning analysis that led to development of a document called the Transportation Planning Report. The report identified ten study corridor options:

- Two options based on improvements to existing US 64, one for the entire length and one for spot improvements throughout the corridor
- Three options on new locations north of the Ocoee River
- Two options on new location to the south of the Ocoee River
- Two options combining new location corridors to the north with improvements to existing US 64
- One no-build option

After additional study, these ten options were simplified to include (1) a no-build option, (2) an option to improve the existing alignment, and (3) several new routes north of the Ocoee River. Two potential alternatives that would have gone south through wilderness areas were eliminated because of environmental concerns. The detailed studies were for a "super two-lane" roadway that included two travel lanes and passing lanes. Concepts for a four-lane highway were dropped because of low projected traffic volumes.

US Highway 93, Montana

US Highway 93, which crosses the Flathead Indian Reservation in western Montana, is often called the Wildlife Highway, or the Peoples Way. Thanks to an unprecedented agreement among the Montana Department of Transportation, the Confederated Salish and Kootenai Tribes (CSKT), and FHWA, guidelines for rebuilding US 93 represented a breakthrough for environmental protection and new alignment of state and tribal interests. With the leadership of landscape architects, the plan opened discussions for creating a more culturally friendly alignment, established an unprecedented number of wildlife crossings, and configured the roadway with more respect for natural

Figure 9-17 It is important for the Corridor K project to have a minimal impact on the region's unspoiled vistas and natural wildlife. Image courtesy of U.S. Forest Service.

Figure 9-18 The Cherokee National Forest is located along the southern ridges of the Appalachian Mountains in Tennessee and provides a wide variety of recreational opportunities. Image courtesy of J. Sipes.

features of the land. New passing lanes and intermittent widening of the highway improve safety. The guidelines were approved in January 2001 and a final master plan was completed shortly thereafter. Construction for most of the highway was initiated in 2003 and completed in 2010. The sections of the highway that go through the Nine Pipe Wetlands Complex are undergoing further study.

A green approach was needed for US 93 because previous efforts to construct the road reached a dead end. In 1990, the Montana DOT proposed to widen fifty-six miles of a principal two-lane arterial highway—US 93—through the Flathead Indian Reservation as a "fast-track project." But after ten years, the project was still stalled.

The Montana DOT goals included improved safety and a higher level of service for US 93, which had an increasing number of accidents. But tribal goals were much broader. Tribal authorities want measures to protect their threatened culture, their sensitive environment, and their breathtaking scenery. The plan to widen US 93 in the traditional way would most likely bring more suburban settlement from the urban areas to the west and to several ancient ecosystems. For the CSKT, this was unacceptable.

Figure 9-19 Although state and federal goals focused on safety while rebuilding US 93, tribal authorities wanted measures to protect their culture, threatened environment, and breathtaking scenery. Image courtesy of Jones & Jones.

Montana DOT and the CSKT were at a standstill. The tribes knew that what the DOT wanted to do was not what they were looking for, but they were unable to articulate what would be an acceptable alternative.

Two consulting firms, Skillings-Connolly, Inc., and Jones & Jones Architects and Landscape Architects, Ltd., were brought in to help develop a collective vision of what this new highway would be. Skillings-Connolly had been working with the CSKT to develop an access plan for the reservation. Jones & Jones was brought in because of successes the firm had on previous context-sensitive highway projects.

Design concepts and guidelines were based upon the premise that the road is a visitor that responds to and is respectful of the spirit of place. Understanding and respecting the spirit of place provided inspiration and guidance that led to design solutions uniquely suited to the special qualities of the tribal lands.

Signed Agreement

On December 20, 2000, FHWA, Montana DOT, and the CSKT signed an historic memorandum of agreement that represented a breakthrough for environmental protection and cooperation between state and tribal interests and moved the project closer to construction.

Figure 9-20 US 93 enters Montana from Idaho, continues north through Flathead National Forest, then reaches its end at the Canadian border. Image courtesy of Jones & Jones.

Design and alignment concepts in the newly adopted guidelines for improvements to US 93 include road alignment, lane configuration, fish and wildlife crossing structures, fencing to complement crossings, interpretive signs and community entry markings, and other roadway features. The guidelines also include standards for limiting and controlling the use of land around the highway. A traffic operational analysis was conducted to ensure that the final design would be the best in terms of safety and level of service. The landscape architect's conceptual alignments and design guidelines were approved.

Wildlife Crossings

Originally, more than fifty wildlife crossings were planned, but that was eventually reduced to forty-two as a result of a value engineering process to reduce cost. This number includes one overpass for wildlife that is landscaped and fenced to provide a linkage zone between wilderness areas to the east and the Cabinet Mountains to the west. The forty-two wildlife crossings are part of a multifaceted strategy to funnel migrating wildlife to safe crossings under and over the road bed. From culverts for fish to bridges and a major underpass, animals will literally see their way to the other side of the road. Where necessary, specially designed fencing was tucked into the surrounding landscape to control animal movement and direct it toward crossings.

The Western Transportation Institute (WTI) at Montana State University conducted preconstruction research from 2002 through 2005 using tracking beds that sampled cross-highway movements for deer, bear, and other wildlife. A small-scale, preliminary post-construction wildlife monitoring project is currently under way as a cooperative study among Montana DOT, FHWA, the CSKT, and the Western Transportation Institute. A small number of motion-sensing animal detection cameras and animal tracking beds have been installed to monitor wildlife crossing structures and their use by wildlife. Through this initial research, wildlife biologists are finding that numerous species are frequently using the crossings.

Figure 9-21 Using natural vegetation for the floor of the wildlife crossings can help make the underpasses more welcoming and more appealing for animals to use. Image courtesy of Jones & Jones.

Figure 9-22 It is sometimes helpful to use fencing to guide the animals toward the opening of the wildlife crossings. Image courtesy of Jones & Jones.

Figure 9-23 This bridge structure is wide enough to allow the creek and its adjacent landscape banks to connect from one side of the road to the other. Image courtesy of Jones & Jones.

Figure 9-24 A scenic overlook allows visitors along US 93 to park their cars and take advantage of views of Flathead Lake and the town of Pablo. Image courtesy of Jones & Jones.

IMPLEMENTATION

A lot has happened with US 93 on the Flathead Indian Reservation since 2000. The majority of the construction was completed by fall 2009. Contract incentives were offered to encourage an aggressive construction schedule, and contractors chose to work into the winter months to finish projects. In addition to roadway developments, other parts of the project include landscape improvements, wildlife crossings, and road enhancements. A visitor center and interpretive overlooks provide opportunities for travelers to rest, relax, and find out more about the native people and the reservation. Thirty-seven interpretive signs and road signs are written in three languages—Kootenai, Salish, and English. The visitor center and overlooks will incorporate the work of tribal artists. In January 2008, the Salish and Kootenai Culture Committee completed work on an audio compact disc about the Salish place name signs on US 93, including pronunciation and translation of all the names.

ACCOLADES

The project has certainly received its share of accolades. When the memorandum of agreement was signed by the three governments, all parties involved celebrated the process and the final decisions. Then Montana Governor Judy Martz said, "The groundbreaking approach to safety and environmental sensitivity of the US 93 project marks a new era in Montana highway construction." Fred Matt, CSKT tribal council chairman, said, "The words in the agreement are about rebuilding a road, but the process leading up to it was about rebuilding trust, honor, and mutual respect among governments." Even the press applauded the project. John Stromnes, writer for the *Missoulian*, wrote in an article headlined "Kinder, Gentler Project" that "This is not your father's highway project." Stromnes also wrote that the project "represents an unprecedented level of environmental protection in road design and a new alignment of state and tribal interests." A Seattle newspaper headline read,

SUNDAY
Missoulian

U.S. Highway 93 expansion

All bark, few bites

Kinder, gentler project

Culture, habitat are centerpieces of expansion across Flathead Reservation

FIGURE 9-25 An article in the *Missoulian* emphasized how culture and habitat are considered the centerpieces of the expansion of US 93 through the Flathead Reservation. Image courtesy J. Sipes.

"Montana highway will redefine environmental sensitivity," and the headline in a Spokane, Washington, paper said, "Project opens new era of harmony between people, habitat and road."

A press release on Montana's official website stated, "When The Peoples Way is completed in 2009, it will be among the most context-sensitive highways in the United States. It will not only reflect the spirit of place, but will allow the rebirth of native grasses, plants, and shrubs along the corridor; the protection of all wildlife living in the Flathead Nation; and the safety of visitors and residents who pass through this land."

Alligator Alley, Florida

Alligator Alley is a section of I-75 and State Road 84 that runs from Naples on the west coast of Florida to Weston, which is near Miami, on the east. Some say the name of the road is a reference to the large number of alligators in the area, but it is reported that the American Automobile Association coined the name Alligator Alley during initial planning efforts because they believed that the road would be useless—an alley just for alligators. The road cuts through a section of the Everglades that includes endangered and unique wildlife, including prime habitat of the endangered Florida panther.

Alligator Alley is often referred to as the most controversial road ever built in Florida because about half the people fought to build the road to connect Weston and Miami to Naples, while the other half fought to preserve the landscapes. The road was first opened in 1969 as a two-lane road.

An environmental impact statement was prepared for Alligator Alley in the 1970s, making it one of the first conducted in the State of Florida.

FIGURE 9-26 Alligator Alley is a section of I-75 that links the Atlantic and Gulf coasts. The roadway was the first of its kind to pass through the Florida Everglades. Image courtesy of Wikipedia Commons.

The road was widened to four lanes between 1986 and 1992 to meet requirements for I-75. One of the biggest concerns during initial planning efforts was the environmental impact that widening the road would have upon the Everglades and the wildlife in the area. As part of this effort, Florida DOT conducted an environmental reevaluation as mandated by FHWA's NEPA regulations and determined that wildlife underpasses and bridge widenings were needed along Alligator Alley to protect important wildlife, including the endangered Florida panther.

The completed project included twenty-four wildlife underpasses, twelve bridge extensions, habitat restoration, and wildlife fencing. Florida DOT purchased land at the State Route 29 interchange to prevent development and helped develop a preserve for the Florida panther. There was also an extensive environmental education effort to let people know about the sensitivity of the environment and the efforts made to protect it, including brochures, informational kiosks, and wildlife warning signs.

Mountains to Sound Greenway, Washington

The concept of green roadways extends beyond the highway right-of-way and includes the surrounding landscape. One of the best examples of this approach is the Mountains to Sound Greenway (mtsgreenway.org/), which surrounds a one hundred–mile stretch of I-90 just east of Seattle, Washington. The greenway passes through protected and working forests, farms, historic sites, lakes, campgrounds, rivers, trails, wildlife habitat, and vibrant communities. It features spectacular alpine scenery, diverse wildlife habitat, world-class recreation, and small, picture-book communities with rich histories of integrating with the natural environment.

FIGURE 9-27 The Mountains to Sound Greenway curves around a forested mountainside in Snoqualmie Pass, with I-90 running through the middle of the greenway. Photo courtesy of Mountains to Sound Greenway Trust, National Scenic Byways Online (www.byways.org).

Motorists driving along the I-90 corridor may mistakenly think that the striking landscape that borders the highway just happens to be that way. The reality, though, is that it has taken almost thirty years of hard work by a variety of interested individuals and organizations to protect the natural resources, prevent strip development along the highway, control urban sprawl, and eliminate billboards and other visual obstacles (Lynch, 2010). The Mountains to Sound Greenway Trust is the group that created a green vision for the I-90 corridor.

The Mountains to Sound Greenway Trust is a nonprofit organization founded by Jim Ellis, Brian Boyle, and Ted Thomsen in 1991 with the goal of protecting landscapes along I-90 and preserving them for public benefit. That year, the three were concerned because they had witnessed so much of King County's farmland and natural areas being converted to residential and commercial uses and didn't want to see the same thing happen to the areas along the I-90 corridor, which were ripe for development. They decided the best approach was to create a permanent greenway along the one-hundred-mile corridor from the Kittitas foothills to Puget Sound (Ellis, 2006).

The stated mission of the greenway trust is to "lead and inspire action to conserve and enhance the landscape from Seattle across the Cascade Mountains to Central Washington, ensuring a long-term balance between people and nature."

In remarks delivered to the Downtown Rotary Club of Seattle on August 30, 2006, Jim Ellis said, "Forests were and are the unifying feature of the Snoqualmie corridor landscape. It was clear from the outset that permanently protecting these forests was going to be Job One for the Greenway Trust."

To achieve greenway forest land goals, public agencies could acquire either fee title to properties or less costly development rights. Land exchanges were worked out with major timber companies as well as with the U.S. Forest Service and the State Department of Natural Resources. They gave up their land within the greenway for land of equal value in other areas. The key to the greenway's success was that it used small land acquisitions to negotiate larger conservation areas by building private and public partnerships. The Greenway Trust's role is not to buy property, but to encourage private and government agencies to do so and to work out a plan to secure that property as public green space (Lynch, 2010).

The Greenway Trust's board of directors and advisory council work together to make decisions about how best to manage the land within the greenway. The board consists of sixty members, including major landowners and managers along I-90, foresters, business representatives, recreation groups, environmentalists, community activists, elected leaders, and representatives of government agencies. The seventy-member technical advisory committee is made up of citizens and individuals with an expertise along the corridor. The technical advisory committee was responsible for developing a comprehensive Greenway Concept Plan that reflects the goals for the greenway.

What makes the Mountains to Sound Greenway so successful is that the Greenway Trust is an inclusive organization that invites all stakeholders to participate in the decision-making process. At a Greenway Trust meeting you will see corporate executives as well as local homeowners, timber company representatives and environmentalists with the Sierra Club, elected officials and volunteers, and land managers with community advocates.

The success of the Mountains to Sound Greenway Trust has drawn the attention of organizations and agencies around the world that seek to duplicate its success. The way the Greenway Trust leverages land acquisitions could be a template for how to build partnerships and take a more holistic approach to protecting natural and cultural resources.

In 1998, the Mountains to Sound Greenway was selected as the first interstate freeway to be designated a National Scenic Byway. The greenway functions as the gateway into Seattle, and it sets the stage for the remarkable visual and natural resources found in the Puget Sound area.

One reason the greenway has achieved so much in such a short period of time is the number of

Figure 9-28 Snoqualmie Point Park has one of the area's greatest views of the Cascade Mountains, Mount Si, and Mount Baker. Image courtesy of Jones & Jones.

volunteers who have donated their time and expertise. Since 1995, the trust has sponsored projects to plant trees, restore stream banks and wetlands, build and maintain trails, and enhance visual resources. In the last ten years these volunteers have planted 500,000 trees, and counting. Since 1995, the trust has offered an environmental education program for King County schools that teaches more than five thousand kids each year (mtsgreenway.org).

Over the last fifteen years, the Greenway Trust has helped broker more than 140 land acquisitions. The greenway forest protection effort has added 130,000 acres to the public lands owned in fee and purchased the development rights for another 90,000 acres to permanently sustain the forests along the corridor. All acquisitions have been from willing sellers. Collectively, the public currently owns about 750,000 acres in the Mountains to Sound Greenway (mtsgreenway.org).

One successful project was the acquisition of land along the crest of Rattlesnake Mountain near North Bend. This part of the mountain is among the most visible in the Cascade Mountains, and there was concern that clear-cutting timber or other impacts would destroy the visual quality of the area. The State Department of Natural Resources, King County, and the U.S. Forest Service combined forces to purchase land from eleven different private landowners, and the result was the Rattlesnake Mountain Scenic Area.

Some of the properties acquired for public use as part of the greenway plan include historic Meadowbrook Farm, a trailhead for hiking trails on Tiger Mountain, Preston Mill on the banks of the Raging River, land around Snoqualmie Falls, Snoqualmie Point Park, and Iron Horse State Park and the John Wayne Pioneer Trail. This last is one of the most popular recreation areas within the greenway. The park and trail are located along a former railroad right-of-way and are managed by the Washington State Parks and Recreation Commission.

Figure 9-29 Park amenities at Snoqualmie Point Park make it ideal for picnics, festivals, and weddings. Included are an open-air amphitheater and an architecturally unique viewing shelter with interpretive markers. Image courtesy of Jim Bennett.

Figure 9-30 The overlook at Snoqualmie Point offers some of the best views of Mt. Si and the I-90 corridor. Photo courtesy of Mountains to Sound Greenway Trust, National Scenic Byways Online (www.byways.org).

Yellowstone National Park's East Entrance Road

The Yellowstone National Park's East Entrance Road was designed by the FHWA Western Federal Lands Highway Division as part of Yellowstone's twenty-year Parkwide Road Improvement Plan. The objective was to improve the entrance road while enhancing the visitor experience and protecting natural and cultural resources. The project includes a 6.9-mile section of road located along a narrow ridge with limited space.

The project involved widening the road corridor to make it safer for the larger number of visitors to the park while also protecting the visual and environmental quality of the existing landscape. The design team included engineers, landscape architects, scientists, and technical specialists from FHWA, the National Park Service, and the Federal Lands Highway Program (FLHP).

In 1982, the Surface Transportation Assistance Act created the FLHP to provide financial resources and technical assistance for public roads on federal and American Indian lands. The Office of Federal Lands Highway oversees the FLHP. The focus of the program is to define the challenges involved in its projects and then propose solutions using innovative, emerging, and underused technologies. The Office of Federal Lands Highway also strives to allow visitors to experience safe and pleasant journeys when driving in national parks, forests, wildlife refuges, and American Indian reservations.

Figure 9-31 The east entrance to Yellowstone National Park climbs over Sylvan Pass and descends to Yellowstone Lake. Image courtesy of FHWA.

For the historic East Entrance Road, the design team sought to minimize the construction footprint and blend the road into the natural environments. New guardrails were constructed of log posts and weathered steel for a more rustic, natural appearance. New blasting techniques were used to sculpt safe rock slopes that resemble natural slopes.

Native vegetation was used to mitigate roadside impacts. A vegetated stone and soil ramp in front of one of the retaining walls helped maintain a movement corridor for bears and other wildlife.

The project used almost 31,000 tons of asphalt; one section of the asphalt used a warm-mix additive, and the remainder was constructed as a control section using traditional hot-mix asphalt. Warm-mix asphalt requires less energy to heat the asphalt, so is more efficient. This was the first time warm-mix asphalt had been used in Yellowstone.

ARC International Design Competition for Wildlife Crossing Infrastructure, Vail Pass, Colorado

The Animal Road Crossing (ARC) International Design Competition for Wildlife Crossing Infrastructure is a yearly competition in which teams tackle new challenges and issues with roadways, structural engineering, and wildlife conservation. The 2010 competition focused on designing a wildlife crossing for I-70 in Colorado.

I-70 is the only east-west road across Colorado, and as a result is heavily used. The proposed location for the wildlife crossing is at Vail Pass, which is located in Eagle County, Colorado, and is surrounded on all sides by White River National Forest. Colorado DOT had previously conducted a study and found that Vail Pass had two areas where the road separated habitats and caused problems for wildlife crossing. The first area did include bridges that spanned drainage areas, allowing animals to cross underneath and avoid the roadway. The second area, however, had no bridge, so wildlife had to walk across the road.

The ARC competition seeks to minimize crashes between motorists and wildlife. In the United States, there are more than 1.6 million deer-vehicle crashes per year. This doesn't include crashes with other wildlife. The collisions with deer result in more than two hundred fatalities and thousands of injuries per year. The average cost to repair a vehicle after an animal-vehicle collision is $2,900, which adds up to more than $8 billion dollars annually. In Vail Pass, the mule deer is only one of many different species whose crossing needs had to be addressed. Clearly something needed to be done to minimize the problem, and previous efforts by the Colorado DOT had not proven successful.

ARC wanted a more effective way to protect wildlife than the types of crossing structures that were being used in other parts of I-70. These crossings were large and bulky and didn't fit well with the surrounding environment. To accomplish this goal, ARC sponsored the competition to develop innovative approaches for a wildlife crossing at Vail Pass.

More than one hundred firms from nine countries combined forces to form thirty-six different teams, and each submitted their qualifications and design ideas to ARC. Of these teams, the ARC selection committee chose five finalists, including Balmori Associates (New York); Janet Rosenberg & Associates (Toronto); Michael Van Valkenburgh & Associates with HNTB (New York); the Olin Studio (Philadelphia); and Zwarts & Jantsma (Amsterdam). These five teams each completed a design that incorporated the stated ARC goals for the Vail Pass crossing. The committee then selected a winner based on a list of criteria, with the main points of interest being innovative design, quality of design, and cost-effective design.

The winner of the contest was HNTB and Michael Van Valkenburgh Associates, Inc. (MVVA). Their design, according to the team, was a "hypar-nature design." The team's basic starting point was that most wildlife crossings are made with humans in mind, and although they help wildlife move across the road, the structures are intended to be aesthetically pleasing to humans. This approach can greatly affect the form of the structure. HNTB and MVVA took a completely

FIGURE 9-32 Olin Partnership's Wild (X)ing includes a bridge with a concave shape with the lowest point in the center of the crossing. Image courtesy of Olin Partnership.

FIGURE 9-34 Wild (X)ing uses a lightweight modular system of rhombic pieces to create the arched bridge to cross the interstate. This model shows how the crossing structure fits with the land on both sides of the highway. Image courtesy of Olin Partnership.

different approach. The form they developed for their crossing structure was determined by the ecological demands of the surrounding environment based upon bands of the landscape (shrubs, forest, meadow, etc.) that were condensed into smaller bands to create a continuous environment for wildlife.

The structure was a precast hyper-parabolic form. This form makes it easier to create, assemble, adapt in the future, and minimize disturbance to the surrounding areas. The adaptability, flexibility, and efficiency of the HNTB and MVVA's hypar-nature design made it appropriate for extensive use. The design team also

FIGURE 9-33 The modular units that make up the Wild (X)ing are linked together and planted with a variety of native plants and trees. The idea is to create a habitat that looks like a natural extension of the landscape. Image courtesy of Olin Partnership.

created a web-based observation platform that allows visitors to see numerous overpasses in real time, as well as data on the amount and types of creatures using the overpass, specific habitats, and normal migration habits.

The ARC International Design Competition for Wildlife Crossing Infrastructure was created to rethink wildlife crossings. The design firm did just that, and as a result they created a new blueprint that allows wildlife and humans to coexist in the same environment. There are no immediate plans to implement the solutions that have come out of the competition, but ARC continues to explore innovative alternatives for addressing wildlife crossings.

Chapter 10

Constructing Green Roadways

The goal of the construction phase of a green roadway is to build with the smallest possible ecological footprint by reducing carbon dioxide, using more recycled materials, reducing the amount of virgin materials used, and lengthening the life span of the road before it has to be repaired or replaced.

New roads can be constructed of recycled tires and shingles, fly ash, reclaimed asphalt pavement, and warm-mix asphalt and other processes that are affordable, energy-efficient, and environmentally friendly. (See "Air Quality" and "Generating Energy" sections in chapter 9.) Part of reducing the ecological footprint is to also take into account the energy used to manufacture materials, transport the materials to a job site, and then build the roadway and supporting transportation structures. For new roads, the production and transport of materials causes significant environmental impacts.

Green Construction Practices

Constructing roads is an expensive, time-consuming process, and the impacts to a community and to the environment can be substantial. Major cities seem to continuously have roads under construction. One of the first steps needed is the development of a construction plan that outlines the process for constructing a given road. The plan defines where materials come from, responsibilities of each team member, processes and standards, and a detailed schedule of tasks. Each plan is unique to address the specific concerns of a project.

Some of the construction practices that can help ensure that a transportation project is implemented in the most sensitive way possible are the following:

Build for the long term. The goal for green roadways should be constructing roads that last as long as possible. Instead of building roads to last fifty years, we should be building them to last one hundred years. Even if initial costs are a little higher, the long-term results will be more cost-effective and will cause fewer impacts.

Schedule construction to minimize impacts. Road construction should be minimized during specific times of years, such as the breeding season for any sensitive species, or during the rainy season, when impacts will be compounded.

Consider the overall energy use of a project. The cost of building a road includes the cost of the energy it takes to deliver materials to a job site. A good rule of thumb is that the energy required to construct a road

Figure 10-1 Ponte dei Salti is a an old stone bridge situated in the Verzasca Valley in Switzerland. It is typically referred to as a Roman bridge, but it was actually constructed in the seventeenth century and was rebuilt in 1960. Image acquired from depositphotos.com.

is roughly equivalent to the energy expended by all traffic driving on that road for up to two years.

Emphasize reuse and recycling. These can significantly contribute to more sustainable road construction practices.

Clear the site only within the limits of construction. Ensure that the project site is cleared only within the limits of construction to avoid excessive site disturbance.

Protect important environmental, landscape, and cultural features. Ensure that trees, shrubs, landscape and cultural features, and environmentally sensitive areas to be preserved are identified and protected during construction. In areas where vegetation is to remain, avoid disturbance and compaction of the ground.

Coordinate with construction personnel when planning and designing projects. Project planning and design requires close cooperation with the personnel directly responsible for construction.

Ensure that erosion and sedimentation are controlled during construction. Ensure that sediments are managed through the timely control of soil erosion. Give preference to sediment control devices, including sediment basins, diversion berms, vegetative buffer areas, channel linings, energy dissipaters, seeding, and mulching. Build permanent erosion controls into structural earthwork design through terracing, flattening slopes, stone and durable synthetic blankets, retaining walls, rip-rap, and/or native revegetation.

Carefully manage and dispose of waste material. Avoid disposing of milled asphalt by placing it as a cover on highway shoulders because this has a negative impact on the visual quality of the roadway. Excess asphalt should be removed.

Salvage and store topsoil and native plant materials. After soil erosion and sediment control measures have been implemented and before grading work begins, remove and store topsoil for project reuse. Salvage areas should be designated on plans and laid out on the site.

Consider location/reclamation of construction areas. Construction staging areas, borrow pits, and other construction areas must be carefully located and restored to a condition as good or better than original.

Figure 10-2 Langkawi Sky Bridge is a curved pedestrian cable-stayed bridge in Malaysia. It is located 2,297 feet above sea level at the peak of Gunung Mat Chinchang. Image courtesy of Wikipedia Commons.

Figure 10-3 The Hoover Dam Bypass Bridge is a four-lane highway bridge that crosses the Colorado River. The bridge, under construction at the time this book was published, links Arizona with Nevada, about sixteen hundred feet downstream from the Hoover Dam. Image courtesy of Wikipedia Commons.

FIGURE 10-4 Many steps and considerations are taken during the construction phase of a project, in which plans that were designed on paper are now taken on-site and implemented on the ground. Image acquired from depositphotos.com.

Green Construction Materials

When it comes to green construction materials, there are three key words to remember: reduce, reuse, and recycle.

Reduce–Reduce the amount and cost of material used, the energy to produce the material, the environmental impact of the material, and waste. Building to last will also reduce maintenance and long-term costs. The lowest life-cycle cost occurs when the surface course is replaced just before it fails and causes damage to the pavement beneath.

New materials such as composites may be part of the solution. Fiber-reinforced polymer composites are lightweight, high-strength, corrosion-resistant materials that can be prefabricated off-site and rapidly installed, minimizing traffic disruption (AASHTO, 2007).

Reuse–Reuse involves removing and processing a material and returning it in place as the same material. Reusing materials is not only cost-effective, it also is environmentally friendly and reduces our use of valuable natural resources. For example, asphalt can be remilled on-site and reused as a new roadway surface.

Recycle–Recycling takes one material manufactured for a specific use and remanufactures that material for a different use. The more we recycle, the more we reduce consumption of resources. When applicable, use reclaimed concrete and asphalt, scrap tires, plastics, steel slag, roofing shingles, coal fly ash, and composted municipal organic wastes. A study in Wisconsin found that using recycled materials in the base and subbase layers of a pavement can result in reductions in global warming potential (20 percent), energy consumption (16 percent), water consumption (11 percent), and hazardous waste generation (11 percent), while extending the service life of the pavement and producing overall life-cycle cost savings of 21 percent. In Europe, some governments have mandated 100 percent reuse of old road-building materials in new road construction.

Concrete, which is the most commonly used material on the planet, is a composite construction material composed primarily of aggregate, cement, and water. Concrete has relatively high compressive strength, so it makes a good surface for roads and other paving. Reinforcing steel is often used to increase the tensile strength of concrete, especially

Figure 10-5 Mechanically stabilized earth panels, featuring concrete textures, are pieced together to form a retaining wall. This construction process helps minimize construction impacts because the panels can be premanufactured and shipped to the site. Image courtesy of Colorado Department of Transportation.

when it is used for bridges, walls, and other structures. Concrete is a popular construction material because it is durable, has a long service life, and is adaptable. As a result, concrete seems to be the paving of choice for roadways in urban areas.

One major problem with concrete is that it is a major contributor to greenhouse gas pollution. Portland cement, which is used to make concrete, emits more than one ton of carbon dioxide for every ton of cement produced, and 6–8 percent of the world's greenhouse gases are attributed to portland cement. Another problem with using concrete is that there is a lot of waste during the construction process. Water is typically used to clean out concrete trucks and to clean off equipment, and the concrete waste causes significant damage to the environment. Additionally, when concrete has to be replaced, it is not biodegradable or environmentally friendly and is typically discarded in landfills. In recent years there has been much more of an effort to recycle concrete. As much as 65 percent of the reinforcing steel in concrete is also recycled after the concrete is crushed. Concrete made with reclaimed concrete aggregate has reduced strength, but is suitable for barriers, pavements, and nonstructural applications.

To reduce concrete's carbon footprint to near zero or less, different approaches are needed. The cement industry has reduced carbon dioxide emissions by using more efficient kilns and processes. Novacem, a British start-up company, is developing cement that does not use carbonates and can make concrete that absorbs carbon dioxide (Fountain, 2009).

Admixtures are often used in concrete mixtures. Admixtures include accelerators that speed up the hardening of concrete, retarders that slow it down, pigments to modify the color, and corrosion inhibitors that minimize the corrosion of reinforcing bars. In addition, inorganic materials such as fly ash, blast furnace slag (a by-product of steel production), recycled plastic, and silica fume (similar to fly ash) are all by-products of industrial processes. The use of these materials lowers material costs and improves strength and permeability, while also reducing energy use and greenhouse gas emissions.

When concrete is used for travel lanes or shoulders, it often has a fairly simple appearance. For plazas, crosswalks, sidewalks, and other uses that require a greater emphasis on visual character, there are several ways to improve the appearance of concrete. These include the following:

Scoring–Scoring refers to using a jointing tool or saw to create indentations in the surface of the concrete.

Patterned concrete–Patterning concrete uses prefabricated dies to create patterns in concrete as it cures.

Finishing techniques–Various textures, such as a broom finish that has a rough appearance, can be applied to concrete surfaces through the finishing process.

Dyes–Dyes can be added to concrete paving to add a touch of color.

The other paving material frequently used is asphalt, which is a sticky, petroleum-based material that serves as the binder for mixing aggregates to create what is commonly called asphalt concrete. Approximately 94 percent of the roads and 85 percent of the parking lots in the United States are constructed of asphalt. There are a couple of advantages of asphalt over concrete in roadways. Asphalt is less expensive, easier and quicker to install, and easier to maintain. Because asphalt can be laid so quickly, and it dries so fast, travel lanes don't have to be closed for very long.

Routine maintenance of asphalt is simple. It involves milling the surface every fifteen to twenty years for recycling, followed by placement of a smooth new overlay. Asphalt is a recyclable material—it can be used over and over, and its life-cycle never ends. Asphalt roads can be dug up and then reused. It is common for reclaimed asphalt pavement (RAP) to be recycled into hot-mix or warm-mix pavement.

Pervious Paving

Porous asphalt, just like pervious concrete, paving stones, and brick pavers, is a pervious material that allows stormwater to percolate and infiltrate to the soil below the paving. Permeable pavements may be constructed as full-depth porous pavements or surface friction courses. Full-depth porous pavements are constructed using specialized asphalt layers or portland cement concrete surfaces that permit water to drain down to a specially constructed crushed stone base. This crushed stone base functions as a temporary stormwater storage area and allows the runoff to infiltrate into the subgrade.

Recycling

Recycling and reusing materials help reduce environmental impacts. For new roads, the production and transport of materials causes significant environmental impacts. Approximately 85 percent of the total energy savings obtained using recycled materials is associated with material production.

Green building projects often divert 95 percent of the construction waste for recycling or reuse. There is no reason that green roadway projects can't do the same. For example, paving such as asphalt and concrete can be recycled so that more than 90 percent of both is being reused. The more we recycle, the less impact on other natural resources.

New materials, construction techniques and designs will enable us to maintain, repair, and replace green roadways more quickly, less expensively, and with longer life spans made possible by improved life-cycle management. The use of local materials helps reduce construction cost and minimize carbon footprint.

The Texas Department of Transportation (DOT) has conducted research in the use of old tires, used copier and printer toner, and fly ash for road paving. They found that all three can be successfully recycled into asphalt, and the result is a pavement that is more durable than standard asphalt (Kelly, n.d.).

Rustic Pavements

Adding a clear polymer resin to gravel can create "rustic pavements" that look like old dirt roads or historic pavements but have the structural capacity to meet modern traffic standards. This material could be used on transportation projects within historic districts or other areas where creating a more rustic look is important. For example, the material could be used in parks, cemeteries, rural areas, or even new communities that are trying to recapture the look and feel of older neighborhoods. It may also be a viable option for multiuse trails.

The Office of Federal Lands Highway (FLH) uses a transparent, amber-colored, synthetic binder that, when mixed with select aggregates, creates a rustic pavement. This mix meets criteria for long-term durability and aesthetics and can be placed using conventional hot-mix asphalt paving techniques. FLH first used this innovative pavement at Virginia's Richmond National Battlefield Park in 2003, and the next year placed it at the Pennsylvania Avenue pedestrian plaza in front of the White House in Washington, D.C.

Making Asphalt Greener

Much of the emphasis on green materials focuses on asphalt products because, of the 2.6 million miles of paved roads in the United States, more than 94 percent have an asphalt surface. In addition, approximately 85 percent of our parking lots are surfaced with asphalt. (See "Air Quality" and "Generating Energy" sections in chapter 9.)

One of the easiest ways to reduce the carbon footprint of roads is to reuse and recycle more asphalt paving. Asphalt has the lowest carbon footprint of any pavement material, so it makes sense to use the material for green roadways. According to the National Asphalt Pavement Association, asphalt pavement is America's most recycled product. As much as 90 percent of asphalt is recycled and reused as RAP. Old pavement is milled from the road, crushed, sent to an asphalt plant, remixed with new aggregate and asphalt binder, and then put back on the road as new asphalt pavement. RAP has been used in this country for more than thirty years, and today, more than one hundred million tons of asphalt paving are recycled every year.

Hot-in-place and cold-in-place asphalt reuse is similar to RAP, but these processes are done in the field. The pavement is milled in place, treated with a binding agent and compacted, and covered with a new asphalt surface. Hot-in-place is used for major surfacing projects; cold-in-place is used primarily for smaller repairs on existing asphalt roadways.

In Boulder, Colorado, 3R Roofing and WorkLife Consulting joined forces to launch the Roofs to Roads project in February 2009. The program recycles asphalt shingles from roofs and uses the materials to create an asphalt mix for repaving roads. Roofs to Roads staff say that 240,000 tons of asphalt shingles are sent to landfills every year from reroofing and demolition projects in Colorado, and twelve million tons nationally, so the potential savings could be substantial.

Warm-mix asphalt was pioneered in Europe more than a decade ago and has started to be adopted in the United States. Some believe that in the next few years warm-mix asphalt will represent 90 percent or more of our production in the United States.

Warm-mix asphalt uses lower temperatures (250 degrees Fahrenheit versus 300 degrees for hot-mix), a foam mixture, and water to coat aggregates with liquid bitumen to make asphalt.

Some transportation departments are looking at ways to recycle old tires and use them for backfill material. Texas DOT found in 2005 that Texans generate twenty-four million scrap tires each year. Nationwide, the number is 281 million discarded tires every year,

FIGURE 10-6 Road construction near the harbor of Victoria in Hong Kong. Image acquired from depositphotos .com.

and that doesn't count the millions of tires already stockpiled. Tire-derived aggregate uses shredded scrap tires to create a low-cost material for road surfaces. Waste tires can be compressed into bales that are used as reinforcement for other backfill materials. Each two-cubic-yard bale uses about one hundred tires, so not only are we creating a useful product, but we are also addressing the discarded tire problem.

Thin whitetopping is the process of rehabilitating distressed asphalt concrete pavements using a concrete overlay. Thin whitetopping has the potential to become a low maintenance, green treatment for asphalt concrete pavements needing rehabilitation in urban areas because of unique advantages offered by concrete.

Perpetual pavement is an asphalt pavement designed and constructed to last more than fifty years without any major maintenance required. The New York DOT has been using a paving composed of three layers of asphalt with a total thickness of nine inches. The primary focus of this approach is to eliminate bottom-up fatigue cracking while still providing a smooth and durable surface.

Making Concrete Greener

Roads with concrete surfaces were originally designed for a twenty-year life, but today more than 60 percent of these roads are thirty years old or older. New concrete pavement designs using dowel bars to transfer loads between panels are being engineered to last fifty years or more.

Portland cement is the most common type of cement used in road building. Unfortunately, it accounts for 6–8 percent worldwide of all emissions of carbon dioxide. Alternative cements allow a reduction in the amount of portland cement used while still producing a high-quality, durable concrete. Fly ash, microsilica, and granulated blast furnace slag can be used to reduce the need for portland cement.

The American Association of State Highway and Transportation Officials' (AASHTO) Materials Reference Laboratory (www.amrl.net) is a great source of information about new construction materials and practices. AASHTO's website, www.transportation.org, contains more than 95,000 pages of information about the transportation world.

Case Studies—Green Construction

New River Gorge Bridge, West Virginia

The New River Gorge Bridge spans a deep gorge, with the New River at the bottom, in southern West Virginia. The new bridge allows US Route 19 to cross the river, reducing what was once a forty-minute drive around one of America's oldest rivers to less than one minute. Construction began in 1974 and was completed three years later in 1977 at a final cost of $37 million.

The New River Gorge Bridge was once the longest steel arch bridge in the world, extending more than 4,224 feet, with the longest single span being 1,700 feet. At a height of 876 feet, it is the second tallest bridge in the United States. In addition, the bridge is noted for its simple yet elegant arch span that helps define the bridge as a visual landmark. In many ways the bridge is a structural work of art that is as striking now as it was when first constructed. Over the years, the unpainted steel used on the bridge has weathered and now has a rustlike appearance that visually fits with the surrounding landscape.

The New River Gorge Bridge is one of the most photographed places in West Virginia. It appears on the state's commemorative quarter, which was released by the U.S. Mint in 2006.

The use of a single span across the gorge minimized environmental impacts that would have resulted from construction of multiple spans. The rugged landscape, with its steep walls, massive boulders, dense vegetation, and exposed cliffs, combines with the bridge to create a strikingly beautiful landscape. There are no piers to affect these natural resources, so the river itself was not compromised.

Figure 10-7 The New River Gorge Bridge in Fayetteville, West Virginia, was helpful in reducing the forty-minute drive down narrow mountain roads and across the river to less than a minute. Photograph by David E. Fattaleh, courtesy of National Scenic Byways Online (www.byways.org).

Figure 10-8 The New River Gorge Bridge is the fifth highest vehicle bridge in the world, sitting 876 feet above the New River. Image courtesy of Wikipedia Commons.

The bridge is located within the New River Gorge National River Park, which is a National Park Service affiliated area. The park encompasses more than seventy thousand acres of land along the New River.

Every year, the bridge is closed to automobiles on the third Saturday of October as part of Bridge Day. The event attracts up to 250,000 visitors a year, including base jumpers, rappellers, and spectators. The impact on the local economy is significant.

Meador Kansas Ellis Trail, Washington State

The Meador Kansas Ellis Trail is located in Bellingham, Washington, a small community about two hours north of Seattle. The trail traverses six blocks in the downtown area and includes the last section needed to complete the Whatcom Creek Trail.

One of the things that makes the trail "green" is its use of Poticrete, a concrete mix made of old porcelain toilets that are crushed and used as an aggregate in the mix. More than four hundred toilets were ground up to create approximately 20 percent of the final concrete mix. Because of the use of Poticrete to construct the trail, it is probably fitting that a commemorative toilet seat has been placed within the paving of the trail. One benefit of this approach is that it minimizes impacts on natural resources by reducing the use of virgin aggregate. Another is that reuse of the toilets diverted about five tons of material from being dumped into local landfills.

In addition to the use of Poticrete, the trail also includes other green elements such as light-emitting diode street lights, porous pavers for pocket parking, recycled concrete aggregate, and asphalt with 30 percent recycled content.

The project was awarded the world's first Greenroads certification. The Greenroads rating system was developed at the University of Washington to help promote sustainable roadways. Work on the rating system started in 2007. After five years of research and development, the Meador Kansas Ellis Trail received the first Greenroads Silver certification, which is awarded to projects that represent a significantly higher level of sustainability than typical road projects. (See chapter 12 for more information on Greenroads and other rating systems.)

The trail project had a budget of $850,000. The ribbon-cutting ceremony for the project was at the end of September 2011.

US 97 Lava Butte-South Century Drive, Oregon

The US 97 Lava Butte-South Century Drive project involved upgrading 3.8 miles of road that runs through the Newberry National Volcanic Monument in central Oregon. The monument, which includes more than fifty thousand acres of lakes, lava flows, and spectacular geologic features, is located within the Deschutes National Forest. Existing roads needed to be upgraded to accommodate the high number of visitors, but any construction had to take into account the unique characteristics of the landscape.

To minimize visual and environmental impacts and improve safety, the design for the project separated northbound and southbound traffic with a forested median that was more than one hundred feet wide in places. The median also helps control access and limit left-hand turns.

Protecting the scenic view was a primary design goal because this segment of US 97 is designated a scenic corridor because it is in the Deschutes National Forest. For example, texture and pigment were applied to bridge abutments to make them blend with the natural geology. The U.S. Forest Service developed a revegetation plan, collected and purchased plants, provided labor and installation to implement the revegetation plan, and is monitoring the results for up to five years.

The number of lanes was increased to accommodate traffic, with two travel lanes for each direction. The existing two-lane alignment was converted into the southbound lanes, and two new northbound lanes were constructed to provide additional capacity.

Figure 10-9 US Route 97 in Oregon is the main corridor east of the Cascade Mountains and serves the population of two major areas: Klamath Falls and Bend. Image courtesy of Oregon Department of Transportation.

This approach reduced costs and impacts because less paving had to be used. The roadway shoulders are constructed to accommodate bicyclists. An interchange at Cottonwood Road was also improved to enable better access.

Wildlife crossing structures and fencing were added to reduce the number of wildlife/vehicle crashes and improve wildlife movement across the road. A primary focus of the wildlife structures was to help restore migratory mule deer herd access to winter range. Each wildlife crossing consists of fencing and two structures, one under each side of the highway. The southern underpass is for wildlife only, but the northern underpass accommodates wildlife when the Lava Lands Visitor Center is closed and is used as a pedestrian crossing when the center is open. The total cost of the fencing and structures is approximately $2.5 million.

An informational kiosk was developed along the road to help inform visitors about environmental education opportunities. A permanent exhibit is being installed at the visitor center that will include photographs, video footage, and educational material about the wildlife undercrossings.

In keeping with concepts of a green roadway, approximately 80 percent of the existing pavement was reused, and RAP accounted for 30 percent of the base course and 20 percent of the wearing course. In addition, more than 90 percent of materials for the project came from within a fifty-mile radius. For longevity, the base thickness of the travel lanes is eleven inches; the use of this type of long-life pavement is expected to decrease life-cycle costs and maintenance.

The US 97 Lava Butte-South Century Drive project also served as the pilot study for the University of Washington's Greenroads rating system. The purpose of the pilot project was to document and collect information about how best to implement the Greenroads process. The process used on this project included three main steps: (1) Greenroads staff talk with the project manager to share information about the project, identify information needs, and clarify what may be needed to obtain credits, or points, to achieve a specific rating. (2) Greenroads staff complete the Greenroads Project Checklist for the project. This is based on the project manager discussion, review of documents provided, and follow-up communication. (3) Greenroads staff generate a report of the results of the case study/pilot project and provide recommendations.

Construction on the US 97 Lava Butte-South Century Drive project was completed in fall 2011 at a cost of $16 million.

Chapter 11

Economics of Green Roadways

At the end of the day, one question always comes up when discussing transportation projects: What will it cost?

In the United States, all levels of government combined spent approximately $3.5 trillion on highways, roads, and local streets between 1945 and 2010. Although that may seem like a decent amount of money, the increase in travel during that same period was so great that most of the capacity and redundancy planned when the system was built has been used up (AASHTO, 2007). Not only do we need to build new roads, we also need to revamp many of our existing roadways and bridges that have experienced years of neglect and are badly in need of repair. Although a very concerning problem, this also means there are numerous opportunities to build new green roadways.

Transportation projects are expensive, and it will be difficult to pay for all of the improvements that need to be made. Funding expected to be allocated for federal and state highways is not even sufficient to maintain existing facilities, much less construct new ones. One significant reason for the current fiscal problems is increased highway construction and right-of-way acquisition costs. The Federal Highway Administration (FHWA) estimated that the highway cost construction index increased 63 percent between 2002 and 2007.

In 2009, the American Society of Civil Engineers estimated that $2.2 trillion had to be invested over five years to bring the nation's transportation infrastructure up to a good condition (ASCE, 2009). The U.S. Government Accountability Office's report, "Determining Performance and Accountability Challenges and High Risks" (2000) concluded that the current federal transportation funding system "poses challenges to introducing performance and accountability for results" into federal highway programs.

In June 2010, Transportation Secretary Ray LaHood said, "In Washington, we all know what needs to be done in transportation. We need to find $500 billion." LaHood suggested that "the federal government needs to look beyond traditional means" to fund a transportation bill (Cassiday, 2010). The year before, President Barack Obama emphasized the importance of shoring up the nation's crumbling infrastructure, but warned that the mounting federal deficit would require "more creative, new approaches to financing" investment in transit, bridges, and road repairs.

One problem is that our total public spending on transport and infrastructure has fallen steadily since the 1960s. In 2011 it was at 2.4 percent of gross domestic product (GDP); in contrast, Europe invests 5 percent of its GDP and China invests 9 percent. This isn't a new problem; the United States has been spending a much smaller percentage of its GDP than European countries for the last fifty years or so. In comparison, in 2006, German road fees brought in 2.6 times the money spent building and maintaining roads there. American road taxes collected at the federal, state, and local level covered just 72 percent of the money spent on highways that year (Economist, 2011).

Funding Road Infrastructure

In the United States, road systems are financed primarily through fuel taxes, vehicle registration fees, property taxes, and sales taxes. The Highway Trust Fund is the primary funding source for federal highway programs. The fund was established in 1956 and is financed primarily through fuel taxes. It was not until 1982 that 20 percent of the fund was allocated for mass transit projects. The federal Highway Trust Fund has traditionally provided the resources for about 45 percent of all transportation construction in this country. But it no longer provides the level of funding needed for expanding and maintaining our transportation system, because the fuel taxes that fund it have not increased.

In 1993, the federal gas tax rate was increased to 18.3 cents for every gallon of gas, and it has not changed since then. The problem is that money from the gas tax doesn't go as far as it used to. Construction costs are much higher than they were in 1993, and alternative fuels, electric cars, improved gas mileage, and increased public transportation means that the amount of gas used per person is going down, resulting in less money from gas taxes.

Currently, the federal Highway Trust Fund is not able to provide the funding that has been promised to states. In 2008, the fund had an $8 billion deficit. This type of deficit is expected to continue, or even increase, until new funding sources are identified.

Typical sources of funding for transportation projects include:

National Highway System–Funds for projects on all National Highway System roadways. The funding split for this program is 80 percent federal funds, 20 percent state funds.

Non-Federal Aid–Funds for construction, reconstruction, and improvement projects on roads and bridges in urban and rural areas at the discretion of the state. The state share is 100 percent of the project costs.

Surface Transportation Program–Funds for projects chosen by states and localities on any roads that are not functionally classified as local or as rural minor collectors. These roads are referred to as Federal-Aid roads. The funding split for this program is 80 percent federal funds, 20 percent state funds.

Congestion Mitigation/Air Quality–Funds for projects in the Clean Air Act nonattainment areas for ozone and carbon monoxide. The funding split for this program is 80 percent federal funds, 20 percent state funds.

Highway Bridge Replacement/Rehabilitation–Funds the replacement or repair of bridges based on structural adequacy, safety, and serviceability. The funding split for this program is 80 percent federal funds, 20 percent state funds.

Interstate Maintenance–Funds rehabilitation, restoration, and resurfacing on the interstate highway system. Also funds the reconstruction of bridges, interchanges, and overpasses along existing interstate routes and the acquisition of rights-of-way. The funding split for this program is 90 percent federal funds, 10 percent state funds.

Federal Aid–Funding for projects that have specialized or proprietary funding or projects for which the specific federal category has not yet been identified.

Transportation Enhancement–Funding opportunities to help expand transportation choices and enhance the transportation experience. This funding is intended for projects that focus on pedestrian and bicycle facilities, educational programs, scenic or historic resources, landscaping, archaeology, rail-to-trail conversion, and cultural resources.

The Recreational Trails Program–This is a federal assistance program of FHWA that provides funds to the states to develop and maintain recreational trails and trail-related facilities.

In addition to these ongoing sources of funding, the American Recovery and Reinvestment Act signed by President Obama in February 2009 provided $26.6 billion for highways, $8 billion for transit, and $9 billion for high-speed rail and Amtrak. Many applauded the move as a long-overdue investment in our transportation infrastructure. The vast majority of the funds were distributed to states, which had to obligate at least half their funding within 120 days (Paniati, 2009).

Federal Funding Process

Federal funds are made available through the following process:

Authorizing legislation–Congress enacts legislation that establishes or continues the existing operation of a federal program or agency, including the amount of money it anticipates to be available to spend or grant to states, metropolitan planning organizations, and transit operators.

Appropriations–Each year, Congress decides on the federal budget for the next fiscal year. As a result of the appropriation process, the amount appropriated to a federal program is often less than the amount authorized for a given year and is the actual amount available to federal agencies to spend or grant.

Apportionment–The distribution of program funds among states and metropolitan areas (for most transit funds) using a formula provided in law is called an apportionment. An apportionment is usually made on the first day of the federal fiscal year for which the funds are authorized. At that time, the funds are available for obligation by a state, in accordance with an approved State Transportation Improvement Program.

Determining eligibility–Only certain projects and activities are eligible to receive federal transportation funding. Criteria depend on the funding source.

Match–Most federal transportation programs require a nonfederal match. State or local governments must contribute some portion of the project cost. This matching level is established by legislation.

Funding Green Roadways

If traditional methods of funding transportation projects no longer work, what new approaches can be used? Meeting America's surface transportation needs will require a multimodal approach that preserves what has been built to date, improves system performance, and adds substantial capacity in highways, transit, freight rail, and intercity passenger rail, and better connections to ports, airports, and border crossings. Meeting several of these multimodal needs will require sources of revenue outside the Highway Trust Fund (AASHTO, 2007).

A number of different approaches are being discussed to provide funding for the next generation of green roadways.

AASHTO Recommendations

The American Association of State Highway and Transportation Officials' (AASHTO) website (www.transportation.org) for Innovative Financing for Surface Transportation provides information on financing and funding alternatives. In 2011, AASHTO recommended a four-phase approach to increasing revenues to the levels needed.

- In phase one, Congress should take action in fiscal years 2009 and 2010 to preserve highway funding at the levels authorized by the Safe, Accountable, Flexible, Efficient Transportation Equity Act: A Legacy for Users (SAFETEA-LU), and avoid cutting the highway program $18 billion from $43 billion to $25 billion, as was proposed.
- In phase two, Congress should restore SAFETEA-LU's purchasing power by increasing highway assistance from $43 billion to $73 billion between 2010 and 2015, and transit assistance from $10.3 billion to $17.3 billion.
- In phase three, from 2015 to 2025, Congress should increase SAFETEA-LU toward meeting the "cost-to-improve" goals, estimated in U.S. Department of Transportation's (DOT) Conditions and Performance Report, but adjusted to year of expenditure dollars using the Consumer Price Index.
- From 2025 and beyond, Congress should use a vehicle-miles-traveled tax to supplement or replace fuel taxes as the principal revenue source for the Highway Trust Fund.

Federal Gas Tax

One of the most talked about options for raising funds for transportation projects is to raise the gasoline tax. This is the current funding mechanism. Currently, it is set at 18.3 cents per gallon. Some states are also considering raising gas taxes.

U.S. gas prices are far lower than those of most other countries, and the average fuel efficiency of our light vehicles is also lower. According to the Energy Information Agency's Short-Term Energy Outlook (Feb. 2008), U.S. gas taxes are less than 10 percent of gas taxes for major European countries. However, given the difficult economic times and the high price of gasoline, there is little chance taxpayers will support an increase in the gas tax. If anything, taxpayers want their taxes lowered, not raised. Then-senators Hillary Clinton and John McCain even promoted a "gas tax holiday" in their campaigns for the presidency. Most economists have ridiculed the idea, saying it would cost the nation billions of dollars and would not lower the cost of gas.

Increasing Efficiency

Transportation departments will have to look at modifying operational budgets to provide savings that may be used for maintenance and construction. When the New Hampshire DOT conducted an audit, they discovered that 62 percent of their highway fund money was used for nonconstruction purposes. About 30 percent of the funds were used to cover the costs of the state police (Sipes, 2008).

Flexible Funding

The basic idea behind "flexible funding" is that federal transportation funds can be used for either highway or transit projects. This is a major change in how transportation policy has controlled funds in the past. This approach allows metropolitan areas to apply federal transportation funds to their highest priority transportation projects.

Bonds

Many states are turning to bonds to make up the difference left by federal funds. In 2007, Texas voters approved Proposition 12, which provided $5 billion for transportation projects. But even that is a drop in the bucket and did not get close to helping the state address its needs.

According to the *Bond Buyer,* municipal bond financing for highway, bridge, and transit projects increased from $17.6 billion in 2001 to $30.6 billion in 2005. In recent years, interest in bonds has continued to grow.

The Build America Bonds program gives bond issuers significant tax credits and federal subsidies for infrastructure investments. During 2010, the program included $118 billion for transportation projects. Build America Bonds were eliminated at the end of 2010, but a new version was proposed in 2011, and it would have a budget of $300 billion to $500 billion. As of May 2012, the program was still being debated.

Transportation Impact Fees

Impact fees are typically one-time charges levied upon new development to pay for new roads or expansion of existing roads. The basic idea is that because private developers benefit financially from their developments, they should incur some of the cost of providing the necessary infrastructure required by these projects. Developers pay impact fees to obtain a construction permit, zoning change, or other permission that allows them to build their project. Impact fees are most effective when the economy is strong because the fees are directly associated with major new construction projects.

Tax Increment Financing

Tax increment financing for transportation is based upon the idea that real estate values are enhanced by the addition of a transit station or new transit service. Tax increment financing is a technique used to create taxing districts that can pledge future tax revenues toward financing development projects, including transportation. It establishes a base-year tax level for a district. Any taxes generated above that base-year amount can be used for transportation projects.

Road Pricing

Road pricing is the practice of charging drivers based upon the costs they impose on others. Many economists have advocated road pricing as an efficient way to reduce congestion, improve the environment, and balance our budget. Road pricing schemes include high occupancy/toll (HOT) lanes; congestion pricing; tolling; and vehicle miles traveled.

One problem with many of the proposals for balancing transportation needs is that they may discriminate against people who can't afford the higher cost. Will we soon get to a point where only the very wealthiest can afford the high registration fees, tolls, and gas prices? Some critics refer to high occupancy/toll lanes as "Lexus lanes" because of the idea that only the rich may benefit from the service (Sipes, 2008).

High Occupancy

High occupancy lanes are the most common type of road pricing used in the United States. One benefit of a high occupancy lane is that it gives motorists the option of whether they want to pay for that service or not.

The concept of high occupancy lanes was developed on I-15 in San Diego, California, in the 1990s. Drivers of single-occupant vehicles were allowed to pay a toll and use an eight-mile stretch of a high occupancy vehicle (HOV) lane. In Orange County, California, variably priced express lanes along the State Route 91 freeway have been in use since 1995. HOV lanes were converted into HOT lanes on I-15 in San Diego, and on I-25 in Denver. The 95 Express Miami Project also converted HOV lanes to HOT lanes.

In May 2008, ten miles of HOT lanes on Washington State Route 167 opened near Seattle between the cities of Renton and Auburn. These lanes are free of charge for buses, motorcycles, and carpools of two or more passengers; single-occupancy vehicles pay a toll collected electronically. In the first two months of operation, the lanes averaged more than one thousand tolled trips per day. The average toll was $1.25, with the toll at rush hour being as much as $9 (Schoen, 2010). If the HOT lane speed gets too slow despite the increased toll price, the HOT lane automatically switches to HOV-only, ensuring that transit stays on schedule.

Virginia HOT lanes projects on the Capital Beltway and I-95/395 include more than seventy miles of highway. In December 2007, the state signed a $1.93 billion contract with a private firm that provided variably priced express lanes. The company paid for the construction and then was reimbursed through tolls. One concern for critics is that the contract is for an eighty-year lease, and that is a very long time for a commitment like this.

Cities such as Los Angeles, New York, and San Francisco are considering converting HOV lanes to HOT lanes.

Congestion Pricing

Congestion pricing involves charging a fee to enter or drive within a congested area. Singapore, London, and Stockholm all have implemented congestion pricing. It is designed to push some less critical or more discretionary rush-hour highway travel to other transportation modes or to off-peak periods. There is a consensus among economists that congestion pricing is one of the most viable and sustainable approaches to reducing traffic congestion in dense areas.

London began using congestion pricing in 2003, and the city has been pleased with the results. Congestion quickly dropped, average traffic speed increased, and hundreds of millions of dollars were raised and invested in public transit. Bus ridership rose dramatically, and use of bicycles also increased.

Tolling

Toll roads have been used in the United States since the 1920s. Tolling is often used to help finance the construction of a new road when other funding is not available. With toll roads, all vehicles have to pay a fee regardless of which lane they use.

Some states already make extensive use of tolls, including Massachusetts, Virginia, and Pennsylvania. In Pennsylvania, the governor proposed selling the Pennsylvania Turnpike, which was the first modern toll road in the nation, to a private sector firm. Florida has used tolling extensively to provide new urban and interurban highways.

In 2008, the city of Chicago signed a $1.83 billion, ninety-nine-year lease with a Spanish-Australian consortium called Cintra/Macquarie to privatize the Skyway commuter bridge. Is this a good deal for the city? Many experts do not think so, because the tolls will continue to increase over the years, and they will never go away. The short-term windfall for the city may turn out to be a nightmare in the long run.

Georgia 400 is a major north-south freeway that opened in the late 1980s and extends north from downtown Atlanta to Coal Mountain. Currently, the road generates more than $20 million in tolls each year, with the tolls originally being scheduled to expire in 2011. It is not a surprise that the Georgia DOT extended the tolls indefinitely to pay for future projects along the road.

In 2005, tolls were collected on 4,600 miles of roads in twenty-five states and generated $7.75 billion in revenue. Expectations are that this number will increase as states look for alternative funding sources. AASHTO estimated that approximately 150 miles of new expressways are constructed each year, and up to 40 percent of these are financed via tolls (AASHTO, 2007).

One major trend with tolling is a shift to all-electronic toll collection with no toll booths or gates that require motorists to stop. One of the most popular approaches is for customers to purchase tags that are attached to the windshield of their car and are embedded with sensors that can be read by monitors on overhead gantries. Customers who choose not to use a transponder are detected at toll zones, an image of their license plate is captured, and they receive a bill in the mail (NCTA, 2008).

There has been a general prohibition on tolling on interstates and other federally assisted highways going all the way back to the 1956 Federal Highway Act, although that has been modified over the years and tolling is now considered on a case-by-case basis.

FIGURE 11-1 The Monroe (North Carolina) Bypass was designed as a single contiguous toll project, meaning that instead of toll booths, the project includes using electronic transponders and cameras so motorists will be able to drive without ever having to stop to pay a toll. Image courtesy of AECOM.

VEHICLE MILES TRAVELED

Charging for vehicle miles traveled involves an electronic user fee system in which vehicles are tracked via a global positioning system. Charges can vary depending upon whether the travel occurred during peak or off-peak times.

Singapore is the first country to implement a vehicle-miles-traveled plan, having done so in September 1998. Starting in 2012, Dutch drivers will pay about seven cents per mile travelled instead of an annual tax.

The Hudson Institute, in its report *2010 and Beyond*, outlined a mileage-based system that is being used in Oregon. The proposal includes not only a fee for vehicle miles traveled, but also an optional fee levied at peak-demand periods (AASHTO, 2007).

Public-Private Funding

With public funds drying up, some states have turned to private investors to maintain existing highways and build new ones. Public-private partnerships are formal collaborations between public agencies and private concessionaires. These partnerships can be an effective way to help implement transportation projects while keeping costs and risks to a minimum.

In May 2012, the City of Chicago unveiled a $7 billion infrastructure plan that uses a nonprofit fund based on private sector resources and experience to address the city's public infrastructure problems. This public/private partnership is intended to rebuild infrastructure without the use of state or federal funds.

In 2009, the state of Indiana signed a $3.85 billion, seventy-five-year lease with private investors in an effort to make up a $1.8 billion shortfall in their road improvements budget. The private investors will maintain the state turnpike and collect tolls.

In New Jersey, lawmakers are debating a proposal to sell a 49 percent stake in the New Jersey Turnpike and Garden State Parkway to private investors (Schoen, 2010).

LOCUS is a national coalition of real estate developers and investors who are involved with transportation projects and smart growth communities. In particular, LOCUS has focused on transit-oriented and mixed-use development because they are effective as economic catalysts. This type of approach makes sense because transportation projects directly impact land use patterns and development opportunities, and vice versa.

One potential problem is that currently private firms can issue tax-exempt bonds to finance infrastructure projects such as airports and seaports, but not highways. Existing federal tax policies will need to change to encourage private firms to invest in transportation infrastructure.

Maintenance and Life-Cycle Costs

It doesn't do any good to build roads if we don't take care of them. Maintenance is a prerequisite for the proper operation of any transportation project. When it comes to transportation maintenance, the old adage "an ounce of prevention is worth a pound of cure" certainly applies.

According to the American Society of Civil Engineers, poor road conditions cost U.S. motorists $67 billion a year in repairs and operating costs—$333 per motorist. In California, Caltrans estimates that they save three to twenty dollars for every one dollar spent on preventive maintenance.

Maintenance Approaches

Some of the ways that transportation departments have attempted to address maintenance issues are as follows:

- Hawaii DOT has implemented an electronic Pavement Management System that helps maintenance workers link trips and plan routes so that they can select the shortest and most efficient route.
- Washington State DOT has a Maintenance Accountability Process tool and field manual to measure and communicate the outcomes of maintenance activities and to link strategic planning, the budget, and maintenance service delivery. Twice a year, field inspections are made of randomly selected sections of highway.
- Oregon DOT Routine Road Maintenance Water Quality and Habitat Guide, commonly referred to within the state as the "Blue Book," provides direction, best management practices, and technical guidance for routine road maintenance activities.
- In 2007, the Virginia DOT implemented a performance measurement framework to monitor road maintenance contracts. The framework consists of five components: level-of-service effectiveness; cost efficiency; timeliness of response; safety procedures; and quality of services.
- Georgia DOT uses NaviGAtor, an intelligent transportation system that monitors more than three hundred miles of highway through the use of more than fifteen hundred video cameras, ninety-seven changeable message signs, and data management strategies that relay real-time traffic information to the state's transportation center.

Maintenance in Northern Climates

Each region has its own unique transportation maintenance issues. In northern climates, salt is frequently used to de-ice the roads. The side effects of salt, though, are that it rusts vehicles and its chloride component can cause environmental pollution. Applying anti-icing chemicals before or at the beginning of a storm event minimizes the compaction of snow so it can be removed more easily.

Transportation departments are trying to minimize the amount of de-icer they use on their roads. The National Cooperative Highway Research Program Report 577, *Guidelines for the Selection of Snow and Ice Control Materials to Mitigate Environmental Impacts,* includes a variety of methods for de-icing. The City of Wheaton, Illinois, began experimenting with a mixture of beet juice and salt called Geomelt in an effort to find a more affordable alternative to using road salt for treating icy winter streets. In Japan, instead of using salt for de-icing, they are trying different materials that would be more environmentally friendly. The Aomori Ecological Recycle Industrial Association in Japan started manufacturing a road de-icing agent that uses scallop shells combined with apple pomace (Navarro, 2007).

In northern states, blowing and drifting snow causes major problems for motorists traveling during the winter. Snow fences create a barrier that results in the accumulation of snow and prevents drifting. According to the National Research Council, it costs three cents to intercept and divert a ton of snow with a snow fence compared to three dollars to plow the same amount of snow. In parts of Iowa, farmers are paid to leave corn rows at the edges of their fields to create living snow fences. The corn rows provide wildlife habitat, reduce soil erosion, and promote no-till agriculture (AASHTO, 2010).

Maintenance Issues in Rural Areas

Many rural jurisdictions are having difficulty keeping up with maintenance of their local and county roads. A big problem is that many rural areas have a limited tax base, and funding is insufficient for building new roads and maintaining existing roads. FHWA has estimated that approximately 40 percent of county roads in rural America are inadequate for current travel, and nearly half the rural bridges longer than 20 feet are currently structurally deficient. State and federal assistance has not provided the support needed to address the issues.

Life-Cycle Cost Analysis

Life-cycle cost analysis (LCCA) is an evaluation technique that supports informed investment decisions. It is a subset of benefit-cost analysis. LCCA provides a comprehensive means to select the most cost-effective alternative for a specific project.

The LCCA approach enables the total cost comparison of competing design alternatives by analyzing all of the relevant costs that occur throughout the life of an alternative. It incorporates discounted long-term agency, user, and other relevant costs over the life of highway, bridge, and other highway assets to identify the best value for investment expenditures. FHWA's *LCCA Primer* provides background for transportation officials to investigate the use of LCCA.

Case Studies—Economics of Green Roadways

Atlanta BeltLine, Georgia

The Atlanta BeltLine Redevelopment Plan is a redevelopment project that integrates land use, transportation, green space, and sustainable development. The BeltLine is made up of a network of parks, trails, and light-rail that follow an historic twenty-two-mile railroad corridor circling downtown Atlanta. The project was actually the brainchild of Ryan Gravel, who came up with the idea of the BeltLine as his joint master's thesis in architecture and city planning from Georgia Tech in 1999. The idea gained momentum, and the basic plans for the BeltLine were approved by the Atlanta City Council in 2005. Sections of the BeltLine are currently under construction.

The BeltLine is listed as a case study in this chapter because the project is one of the most comprehensive economic redevelopments ever undertaken in Atlanta. A series of urban development nodes is planned for key intersections where major roads cross the BeltLine. These nodes will be mixed-use urban centers that include commercial, retail, residential, and entertainment sites. Eventually each node is intended to be connected by light-rail that follows the old railroad line, but this will take years to complete.

Over the next twenty-five years, the BeltLine is expected to result in nearly thirteen hundred acres of new parks that will almost double the amount of green space within the city. These new parks will be connected with approximately seven hundred acres of existing parks, including Piedmont Park, which is Atlanta's signature park.

The BeltLine is a partnership between the City of Atlanta and the Atlanta BeltLine, Inc. In addition, the

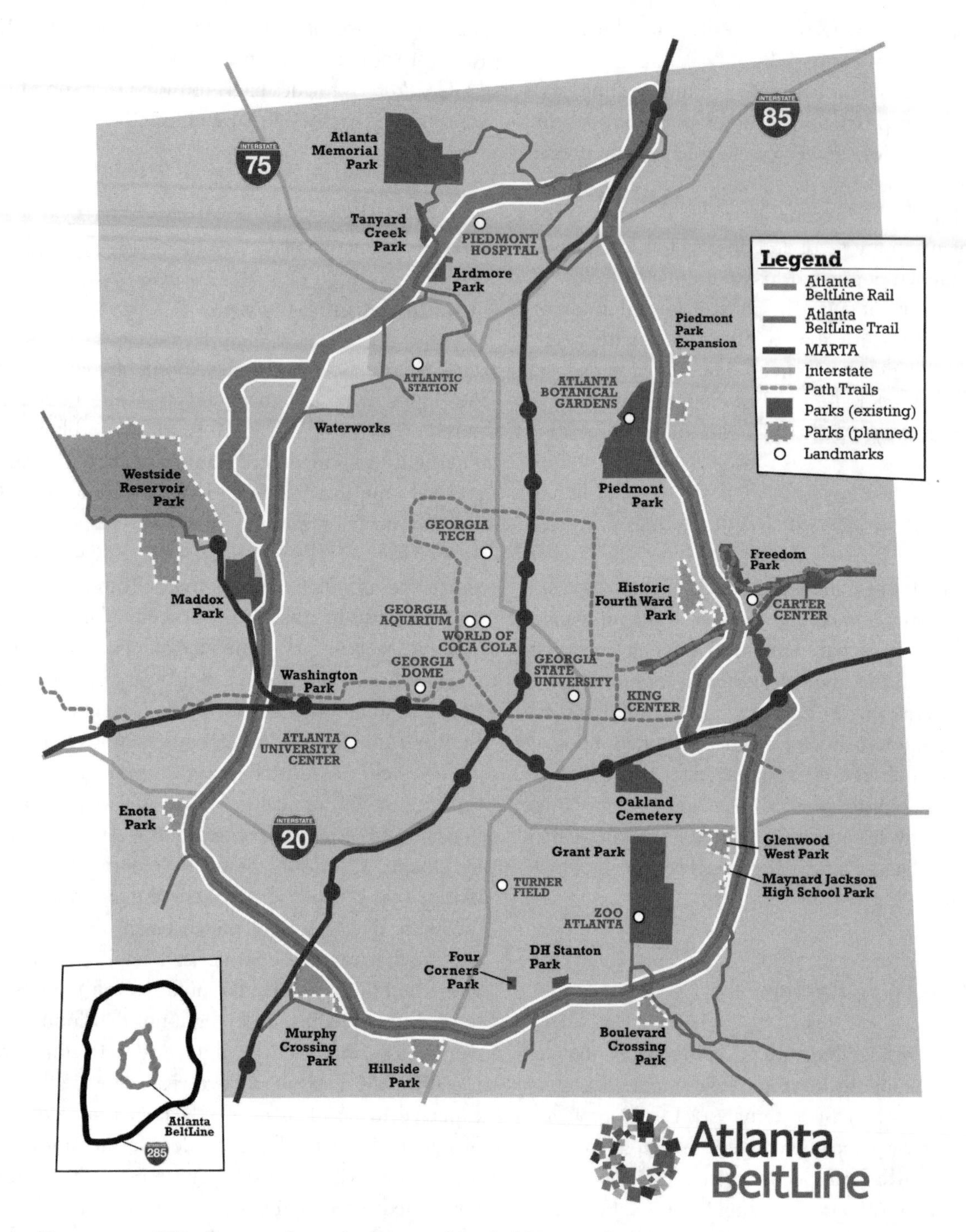

FIGURE 11-2 This diagram shows the layout of the BeltLine, including its connection to existing and proposed green space. Image courtesy of EDAW.

FIGURE 11-3 Development nodes along the BeltLine would provide opportunities for economic growth. Image courtesy of AECOM.

FIGURE 11-4 The addition of pedestrian trails and light-rail as part of a linear park circling the city will help reduce traffic demands in Atlanta. Image courtesy of EDAW.

Atlanta BeltLine Partnership, a nonprofit organization, is involved with raising funds to support the BeltLine. The organization also works with stakeholders to raise awareness and support for the BeltLine. Because the project will take so long to implement, it is important that the public be supportive of the BeltLine during each phase of planning and construction. Other partners for the BeltLine include the PATH Foundation, The Trust for Public Land, the Metropolitan Atlanta Rapid Transit Authority, and many private, public, and nonprofit organizations.

The BeltLine is expected to cost somewhere around $2.8 billion once it is completed in twenty-five years. Some of the funding options being used for the BeltLine include:

Tax allocation district (TAD) financing–A tax allocation district was set up to provide the primary source of funding for the BeltLine. Over the life of the project, the district is expected to generate up to $1.7 billion.

Capital campaign–The BeltLine Partnership initiated a $60 million capital campaign to raise funds for initial infrastructure development.

FIGURE 11-5 The seventeen-acre Fort Ward Park, which is one of the first projects constructed as part of the BeltLine, integrates stormwater detention as part of public open space. Image courtesy of EDAW.

Federal funding–Federal funds are being pursued for eligible projects along the BeltLine.

City of Atlanta funding–The City of Atlanta has invested approximately $165 million in the BeltLine as of 2011 through Park Opportunity Bonds and Capital Improvement Program funds.

Upon completion, the BeltLine is projected to generate more than $20 billion of new economic development and approximately thirty thousand new jobs.

Greater East End Livable Centers, Houston, Texas

Houston's East End District was the birthplace of Houston. The district includes more than four hundred acres located east of the downtown area. More than half a billion dollars is being invested in the district by private developers and through city, state, and federal funds as part of efforts to create a more walkable community. This construction includes the Harrisburg Light-Rail Line, which extends through the heart of the district, and major street improvements that are significantly changing the visual character of the neighborhoods by adding pavers, trees, pedestrian lighting, street furniture, trails, and other site amenities.

The Greater East End is a neighborhood in transition from an industrial/warehouse character to a mixed-use urban neighborhood. There are hundreds of acres of vacant, underused, or obsolete industrial land that are actively being assembled by developers, and a number of new projects have already been constructed. Because of its proximity to downtown Houston, the Greater East End is expected to be one of the next major growth areas in Houston. New infrastructure improvements are needed for the district to develop as planned. This includes reinforcing the existing street network, developing new links and connections, and transforming Navigation Boulevard into a traditional main street.

The Greater East End Livable Centers project was developed through the Houston-Galveston Area Council's (H-GAC) Livable Centers program. H-GAC is the regionwide voluntary association of local governments in the thirteen-county Gulf Coast planning region of Texas. Its service area is 12,500 square miles and contains more than six million people. H-GAC works with local governments to develop "livable centers," which are walkable, mixed-use places that provide multimodal transportation options, improve environmental quality, and promote economic development.

FIGURE 11-6 This image, from Harrisburg at Lockwood with downtown Houston in the background, shows a vision of the East End with the proposed light-rail operational. Image courtesy of Greater East End Management District.

FIGURE 11-7 Finished sidewalks, trees, and additional enhancements were made possible by a partnership between Greater East End Management District and the Metropolitan Transit Authority Image courtesy of GEEMD.

FIGURE 11-8 Streetscape improvements have made the East End more appealing to cyclists and pedestrians. Image courtesy of GEEMD.

The first stage of the Livable Centers Study was to develop a collective vision and priorities for future development of the district. A series of stakeholder meetings, public workshops, and a planning charrette were conducted toward the end of 2010 and beginning of 2011, and out of that came the vision of the district as a "vibrant, mixed-use, multicultural, and sustainable model for the redevelopment of a historic and strategically located Houston neighborhood."

The East End plan is based upon principles that can be grouped into three general topic areas: those that deal with improvements to the street network and traffic pattern, those that promote sustainable development, and those that are concerned with neighborhood character. The district has been able to leverage the light-rail construction with a $5 million American Recovery and Reinvestment Act stimulus grant to implement streetscape and pedestrian/bicycle mobility improvements to three key corridors: Navigation Boulevard and York and Sampson Streets.

The district's vision calls for economic growth by attracting, building, and promoting new development. Economic programs for the Greater East End Management District (GEEMD) focus on six categories: affordable housing, environmental programs, general business assistance and incentives, historic preservation, infrastructure and public facilities, and labor force and training.

Capital improvements along the Atlanta Belt-Line included new sidewalks, lighting, landscaping, benches, and way-finding features. GEEMD is also exploring the use of streetcars to connect neighborhoods to planned light-rail. Streetcars would connect the light-rail to local residential areas, allowing local residents to travel the last mile or two from the light-rail to their destination without needing a car at all.

Chapter 12

Next Steps in Creating Green Roadways

How do we plan for the next generation of green roadways? Forecasts indicate that the U.S. population will grow from 300 million today to 435 million by 2055. Annual highway travel measured in vehicle miles traveled may increase from three trillion today to as much as seven trillion by 2055 (AASHTO, 2007).

In 2005, Congress created the National Surface Transportation Policy and Revenue Study Commission and directed it to develop a "conceptual plan" to address transportation in the United States over the next thirty years. Two years later, the commission published *Transportation for Tomorrow: Report of the National Surface Transportation Policy and Revenue Study Commission*, which includes detailed recommendations for creating and sustaining a surface transportation system in the United States. The commission expired in July 2008. The report states, "The American Transportation Network of highways, transit, rail, and ports is poised on the threshold of a period of innovation unprecedented in our history. The benefits from forward-looking investment will be the underpinnings of a thriving national economy, maintaining America as the international leader in technology and wealth creation, with benefits flowing to all citizens" (AASHTO, 2007).

Four strategic actions can bring about this change: preserve the current system, enhance its performance, expand capacity to meet future needs, and reduce growth in highway demand by expanding the capacity of transit and rail.

Building upon those ideas, we wanted to define our recommendations for the next steps in creating green roadways. These start with embracing the National Environmental Policy Act and emphasizing the concept of evaluating sustainability based on social, economic, and environmental impacts. All three are important in creating green roadways. It is possible to build new roads and expand existing ones while minimizing impacts on environmental resources. Where impacts can't be avoided, design solutions can minimize problems, and innovative solutions can create green roadways that "fit" their surroundings.

Planning for Green Roadways

Creating green roadways starts with a comprehensive, holistic planning process that defines a clear vision for a specific project. Some of the recommendations for a green planning process include the following:

Use an integrated approach–The process of designing green roadways should ensure that community values; natural, historic, and cultural resources; and

FIGURE 12-1 The Sound Tube that is part of the Tullamarine Freeway in Melbourne, Victoria, Australia, is used to reduce noise pollution to neighboring residential development. Image courtesy of Wikipedia Commons.

transportation needs are fully considered throughout the planning, design, and construction phases of a project.

Get the public involved–Transportation projects have a major impact upon the character of a community, and local residents should have a say in such projects. The earlier the involvement, the better.

Be more proactive–We need to be more proactive in rebuilding our infrastructure in a green manner. The Urban Land Institute's *Infrastructure 2007: A Global Perspective* stated that "America is more of a follower and no longer a world leader when it comes to infrastructure." That needs to change.

Expand research–Advanced technologies can help us design, build, and manage our roads more efficiently. Intelligent transportation systems can be used to monitor traffic, the condition of bridges and roads, and weather information. Improved asphalt pavement technologies need to be developed that last longer, are easy to repair, cost less, and are recyclable.

Figure 12-2 Short-term and long-term environmental impacts need to be considered for all transportation projects. Image acquired from depositphotos.com.

Figure 12-3 Effective transportation planning is an integral part of efficient energy use. Image courtesy of J. Sipes.

Figure 12-4 The Gateshead Millennium Bridge, a pedestrian bridge that spans the River Tyne in England, is designed to tilt up to allow river traffic to pass. Image courtesy of Wikipedia Commons.

Fix what we have–The first priority should be to repair and modernize our existing system of highways, transit, and rail. A good starting point is to repair the bridges and roads that are unsafe.

Make our roads safer–Although roadways are safer than they once were, car crashes are still the leading cause of death for persons between the ages of three and thirty-three. The next generation of green roadways could include collision avoidance systems and other technologies that reduce accidents.

Changing the Emphasis of Transportation Funding

Increase alternative transportation–There needs to be a stronger commitment to transit, biking, and other alternative means of transportation. The answer is not simply building more roads. For transit, many bus, light-rail, subway, and commuter-rail systems need to be modernized and expanded.

Get people out of their cars–Create walkable communities that use trails, sidewalks, linear parks, and public gathering spaces, and encourage people to park their cars and walk, run, roller blade, and cycle. The Federal Highway Administration (FHWA) forecasts that highway travel will increase at a rate of 2.07 percent per year through 2022. If this rate continues, highway vehicle miles traveled will more than double from three trillion today to nearly seven trillion by 2055. Getting people out of their cars will make these numbers more manageable.

Develop more high-speed rail–A new National Rail Transportation Policy must be established to increase freight rail capacity and intercity passenger rail services.

Invest more funding–China spends 9 percent of its gross domestic product on infrastructure, India spends 3.5 percent, and the United States spends 0.93 percent. We need to invest more money in green infrastructure.

Provide new funding sources–We have to find new ways to fund transportation projects. Today, the federal government funds about 45 percent of transportation projects, and state and local governments

FIGURE 12-5 Photoelectric panels can be located along highway rights-of-way and used to generate energy, both for highway uses and for surrounding development. Image courtesy of J. Sipes.

provide the remaining 55 percent. To provide the funding needed to maintain and expand our existing transportation infrastructure, new financial resources are required. Public/private partnerships need to become a bigger part of the mix. In *Transportation—Invest in Our Future,* AASHTO officials recommend that three new sources of funding should be authorized to help keep America competitive in the world economy. They are critical commerce corridors, tax credit bonds, and investment tax credits.

- A critical commerce corridor focuses on the creation of an integrated, national strategy for safe and efficient movement of freight and a reduction of the impact of truck traffic on other highway users. The initiative would be financed with dedicated and protected user fees levied on freight shipments.
- Tax credit bonds may be issued by state and local governments and governmental entities for a wide array of purposes, including green roadway projects. These bonds basically provide an interest-free loan to the issuer. The Build America Bonds are an example of a tax credit bond.
- Investment tax credits allow companies to deduct a percentage of investment costs from their tax liability in addition to the normal allowances for depreciation.

Use congestion pricing–Higher fees should be charged during peak travel periods to raise funds and encourage a change in driving patterns.

Measuring Success for Green Roadways

To determine how successful we are at creating green transportation projects, we have to be able to measure success. Several agencies and organizations have developed scorecards to help make the best design and planning decisions when it comes to green roadway projects.

LEED-ND Guidelines

The U.S. Green Building Council's (USGBC) Leadership in Energy and Environmental Design (LEED) Green Building Rating System is the nationally accepted benchmark for the design, construction, and operation of high performance green buildings. The council's LEED for Neighborhood Development (LEED-ND) is a rating system for neighborhood location and design that combines the principles of smart growth, urbanism, and green building.

The current draft of the LEED-ND rating system is expected to undergo significant revisions before it is used for the LEED-ND pilot program. To qualify for LEED-ND certification, a project must meet minimum performance standards. Based on the total number of points received, a project can be awarded certification at one of four levels: Certified, Silver, Gold, or Platinum. Prerequisites and credits are divided among four sections, based on the nature of the associated performance standard: (1) smart location and linkage, (2) neighborhood pattern and design, (3) green construction and technology, and (4) innovation and design process.

Sustainable Highways

Self-Evaluation Tool

FHWA's Sustainable Highways Self-Evaluation Tool is intended to help state and local transportation agencies incorporate sustainability best practices into highway and other roadway projects. The Sustainable Highways Self-Evaluation Tool is a collection of best practices that agencies can use for self-evaluation of programs to determine a sustainability score in three categories: (1) system planning, (2) project development, and (3) operations and maintenance. This tool is designed to be applied to all roadway projects.

Zofnass Program

Harvard University's Zofnass Program is a framework for evaluating infrastructure sustainability. The objective of the program is to develop a set of recommended guidelines for sustainable large-scale projects, including infrastructure projects.

The Zofnass Program focuses on quantitative methods that can be used for measuring sustainability. The first level of research provided a systematic review of sustainable design and development metrics and methods. The second level of research presented case studies that focus on current sustainability practices and methods applied in infrastructure and large-scale projects. The third level of research integrated findings from the two previous levels of research to produce sustainability standards. The Rating System Manual was made available for download in March 2012.

ASCE/ACEC/APWA

The American Council of Engineering Companies (ACEC), the American Society of Civil Engineers (ASCE), and the American Public Works Association (APWA) are jointly developing a sustainable infrastructure project rating system. The goal of the system is to enhance the sustainability of the nation's civil infrastructure, including transportation, water, and environmental projects.

Under the program, sustainable infrastructure project ratings will be recognized only after an independently verified performance assessment. This will give the program a higher level of credibility than programs that are self-verified.

Greenroads

Greenroads is a rating system for sustainable road design and construction developed by researchers at the University of Washington in partnership with CH2M Hill. Greenroads is intended to be applicable to all roadway projects, including new, reconstructed, and rehabilitated roadways. It offers incentives that encourage users to conserve resources, reduce waste, and make longer-term choices.

The voluntary measures in Greenroads fall into five categories: environment and water, access and equity, construction activities, materials and resources, and pavement technologies.

BE2ST in-Highways

Building Environmentally and Economically Sustainable Transportation Infrastructure-Highways (BE2ST in-Highways) is a green highway construction rating system. BE2ST in-Highways employs life-cycle analysis techniques to provide an overall assessment of the environmental impacts associated with a highway construction project.

GreenLITES

The New York State Department of Transportation's (DOT) GreenLITES is geared toward improving maintenance and operations. It uses an annual maintenance and operations plan, which is a comprehensive management system that is used to plan, fund, track, and rate operations activities. New York State DOT developed the GreenLITES certification program to better integrate sustainable transportation principles.

GreenLITES identifies more than 175 sustainable items in five categories: (1) sustainable sites, (2) water quality, (3) materials and resources, (4) energy and atmosphere, and (5) innovation/unlisted.

I-LAST

I-LAST (Illinois–Livable and Sustainable Transportation) is a sustainability guide and rating system developed by the Illinois DOT, the American Consulting Engineers Council–Illinois Chapter, and the Illinois Road and Transportation Builders Association.

The I-LAST guide consists of more than 150 possible sustainable or livable practices that may be included in highway projects. These practices include design-phase activities, such as using context-sensitive solutions practices; design decisions, such as alignments to avoid environmentally sensitive areas or inclusion of transit facilities; and construction specifications, such as allowing the reuse of reclaimed materials.

Sustainable Sites Initiative

The Sustainable Sites Initiative (SITES) is an interdisciplinary effort by the American Society of Landscape Architects, the Lady Bird Johnson Wildflower Center at The University of Texas at Austin, and the U.S. Botanic Garden. SITES was created to promote sustainable land development and management practices that can apply to sites with and without buildings.

SITES is similar to the LEED-ND rating system in that they both significantly extend the focus of green building beyond the single-building envelope. The difference between the two is that SITES focuses on the site scale and LEED-ND focuses on location and community pattern. The two systems are intended to be complementary each other.

U.S. Case Studies—Next Steps

Manchaca GreenWay, Austin, Texas

The Green Mobility Challenge, hosted by the Texas DOT and the Central Texas Regional Mobility Authority, challenged highway engineers, planners, architects and landscape designers to find innovative and sustainable ways to build, operate, and maintain the Austin region's transportation system. The Green Mobility Challenge was intended to represent a bold and innovative step in the City of Austin's green evolution. In the years ahead, the city plans to construct mobility improvements in some of the corridors that have politically divided the community for decades.

One of the two projects in the Green Mobility Challenge was the Manchaca Expressway, a proposed 3.6-mile controlled-access roadway in southwest Austin. The winning team from AECOM Technology Corporation was recognized for their innovative ideas, which included building a contiguous park with a shared-use path along the entire corridor, using state-of-the-art intersection design to eliminate traffic signals, using green shoulders and biofiltration to protect the Edwards Aquifer, and building special roadway crossings to serve as wildlife corridors.

The focus of the design for the Manchaca GreenWay—"greenway" being a much more appropriate description than "expressway"—was about shifting the road's mission from that of a single-purpose spine to that of a multipurpose network of connectivity opportunities for communities, people, and wildlife. It was also about connecting different parts of Austin via the roadway while also promoting multimodal connectivity.

The roadway is designed to fit the land by using a very wide median that allows each travel lane to be adjusted independently. The medians serve as bioswales that handle all of the stormwater runoff for the road. Runoff is treated to remove total suspended solids. The medians look like dry creek beds, with gravel beds meandering through the middle. Curvilinear native plant massings are intermixed with the dry creek beds to add visual interest while reducing maintenance. Fields of native wildflowers located along the edges of the corridor flow across the median, creating a tapestry of color and texture that visually creates a sense of unity.

A double teardrop interchange is a unique double roundabout configuration. It increases capacity with fewer approach lanes, and reduces accidents by a staggering 78 percent while also improving traffic flow and slowing down traffic. The double teardrop intersection uses one-third less area than a traditional

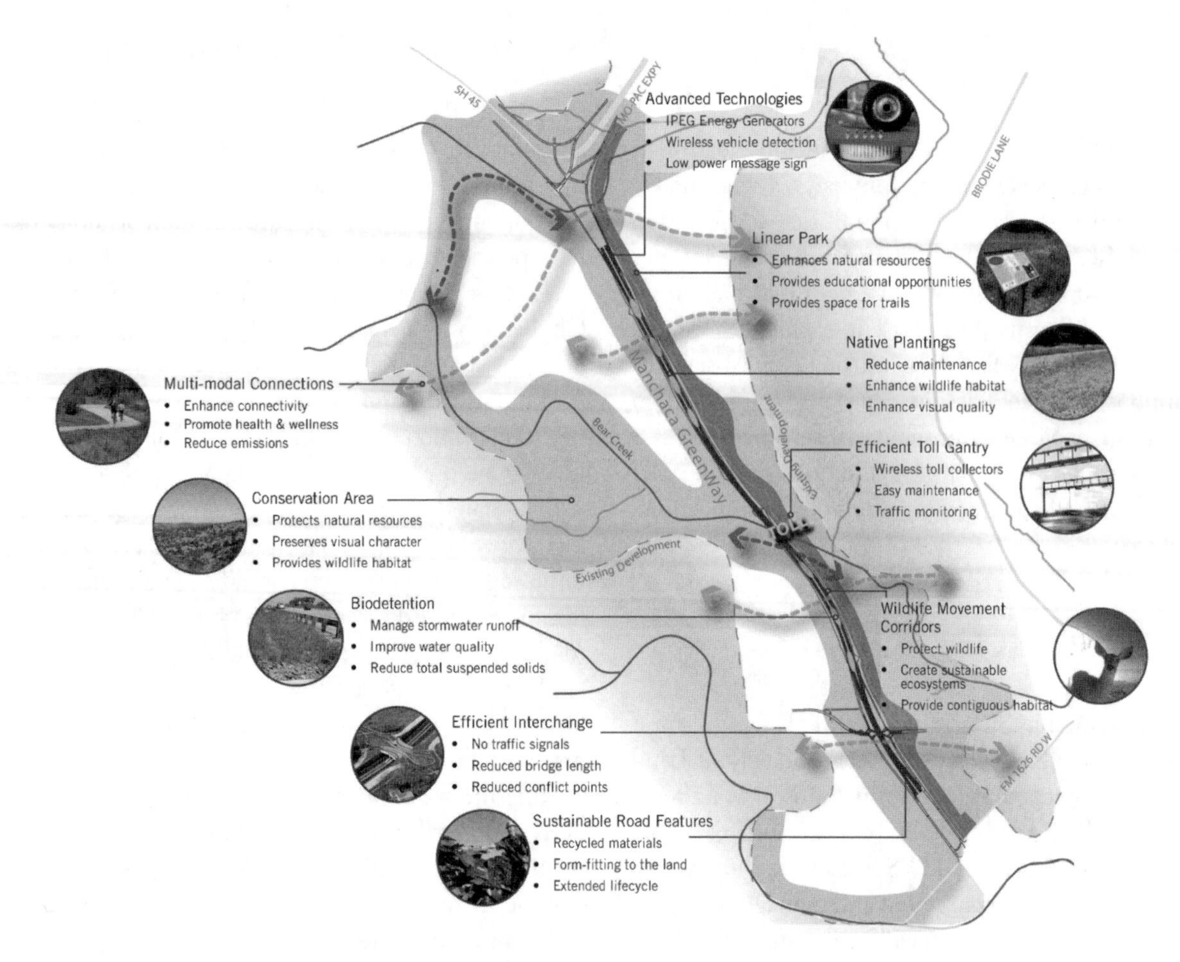

FIGURE 12-6 The proposed Manchaca GreenWay in Austin, Texas, is developed around the concepts of connectivity, protection, and enhancement. Image courtesy of AECOM.

FIGURE 12-7 The Manchaca GreenWay consists of travel lanes separated by a forty-eight-foot median that includes a bioswale and native plants. A linear park and conservation open space help create a green corridor. Image courtesy of AECOM.

diamond intersection, preserving existing vegetation and minimizing impacts.

The GreenWay takes an integrated stormwater management approach to control the volume and velocity of stormwater runoff, filter the runoff to address water quality, maintain normal stream flow in Bear Creek, continue to nourish surrounding ecosystems, and protect the existing Edwards Aquifer. Water treatment elements include porous friction course, green shoulders, bioslopes, vegetated swales, modular biofiltration, vegetated buffers, and biodetention areas.

A conservation area is an essential part of the Manchaca GreenWay because it helps preserve the sense of place that makes this area unique. The GreenWay also include two parks: a linear park that runs along the roadway and a park at the interchange that serves as a trailhead and park-and-ride while providing additional recreational opportunities. The linear park serves as a buffer to protect existing natural resources, helps manage stormwater runoff, and provides the necessary space for trails.

One major focus of proposed innovative technologies is to create an entirely off-the-grid roadway. Reducing dependence on grid power eliminates a lifetime of energy use and carbon dioxide emissions. The roadways of tomorrow can be used to generate electricity on-site, and the GreenWay does just that by using solar panels and technology such as the Innowattech Piezo Electric Generator. This technology harvests wasted mechanical energy imparted into the pavement by converting mechanical stress into current or voltage through a series of in-roadway piezoelectric generators. Installed in the free lanes approaching the toll gantry, this system would be used to power the electronic toll collectors. In addition, modular wireless electronic toll collectors are used for on-demand toll collection and traffic monitoring.

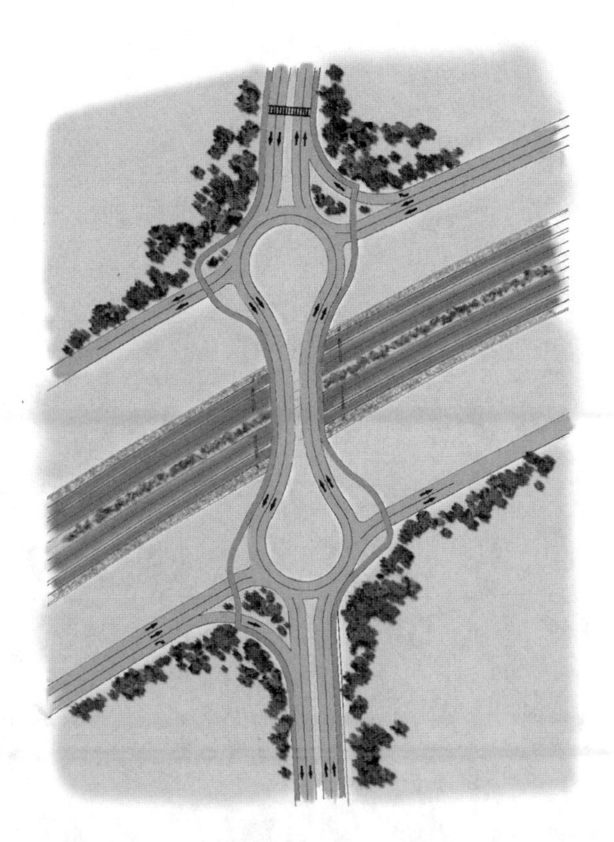

Figure 12-8 Trails run along both sides of the double teardrop intersection of the GreenWay, connecting to multiuse trails running parallel to the highway. Image courtesy of AECOM.

Oregon Solar Highway

The Oregon Solar Highway is the first solar highway project in the nation. In February 2008, the Oregon Transportation Commission approved development of solar installations on Oregon DOT properties, including within the rights-of-way.

The idea of generating electricity from solar panels along highway rights-of-way has been of great interest to FHWA and the U.S. Department of Energy for years. Unfortunately, it seems that going from concept to reality has proven to be a major obstacle until recently. Oregon DOT was able to develop a plan to not only implement solar panels within the right-of-way, but also to make it cost-effective.

On December 19, 2008, Oregon DOT's first solar installation started feeding clean, renewable energy into the electricity grid. The project consists of 594 solar panels and is situated at the interchange of I-5

FIGURE 12-9 The conservation area for the GreenWay works in combination with regulatory incentives and mandatory headwater setbacks to preserve sensitive environmental features that are part of the Texas Hill Country. Image courtesy of Wikipedia Commons.

and I-205 south of Portland, Oregon. The installation, a 104-kilowatt (DC) ground-mounted solar array, is located within the highway right-of-way. It produces almost 130,000 kilowatt hours annually. Solar energy produced by the array feeds into the grid during the day, in effect running the meter backward for energy needed at night to light the interchange through a solar power purchase agreement with Portland General Electric.

The project is owned and operated by SunWay1, LLC, and managed by Portland General Electric, the utility serving the area. The prototype project cost $1.28 million, but Oregon DOT did not have to front the cost because of the partnership agreement with SunWay1. The project was financed through SunWay1 using the state's 50 percent business energy tax credit, the 30 percent federal investment tax credit, accelerated depreciation, and utility incentives. Because SunWay1 has private ownership, it could take advantage of financing opportunities that would not be available to Oregon DOT.

The project demonstrates that solar arrays can complement and not compromise the transportation system, and Oregon DOT has plans for continuing

FIGURE 12-10 The solar panels run parallel to the highway near the interchange of I-5 and I-205, and are located so their visual impact is limited. Image courtesy of Oregon Department of Transportation.

to expand the idea of the solar highway. The agency plans to expand the use of roadside solar to provide the electricity needed to run the state's transportation system.

Houston Low-Impact Development–Independence Parkway, Texas

In fall 2009, the Houston Land and Water Sustainability Forum sponsored a competition designed to explore the benefits of low-impact development (LID) practices. The challenge was to design a new green roadway section that incorporates LID techniques, promotes infiltration, reduces stormwater pollution, and reduces long-term maintenance costs. Due to the Houston region's high-intensity rain events and dense clayey soils with minimal infiltration capacity, it is often assumed that LID would not be applicable or useful in the Houston region.

A one-mile stretch of Houston's Independence Parkway was selected as the site for the competition. Each design team was asked to develop a plan that would widen the parkway from two to four lanes. AECOM's winning entry moved the paved roadway sections to the outside edge of the right-of-way, creating a forty-eight-foot-wide center median slanted inward to collect stormwater. The design has an inward-sloping roadway with continuous curbs on the outside and slotted curbs on the inside. Bioswales slow down and clean the water as it passes through the system. The plan results in significant reductions in highway

FIGURE 12-11 The 594 solar panels located with the right-of-way of the Oregon Solar highway produce nearly 130,000 kilowatt hours of electricity annually. Image courtesy of Oregon Department of Transportation.

FIGURE 12-12 The green cross section of the Houston LID project includes an inward-sloping roadway with continuous curbs on the outside and slotted curbs on the inside to convey roadway runoff into a forty-eight-foot-wide center median bioswale. Image courtesy of AECOM.

runoff pollutants, removing 84 percent of total suspended solids, 68 percent of metals, and 30 percent of pathogens.

The landscaping consists of strong, dynamic patterns that address both form and function. The design team wanted to create a landscape that was affordable, easy to care for, and environmentally sustainable. Gravel ribbons create the appearance of dry creek beds, and ornamental plant massings add visual interest. In selected areas, rainwater harvesting areas are used to capture stormwater, either allowing the water to percolate into the soil or to be directed into an underground cistern. To create a more sustainable solution, the plan uses underground cisterns, solar-powered drip irrigation, and bioharvesting within the right-of-way. Alternatives such as switchgrass harvesting in expanded rights-of-way could create additional functional spaces that would improve aesthetics along the roadway and which, through biomass harvesting, could provide a potential new source of revenue for the county. Despite all of these improvements, the green roadway design costs 13 percent less than the standard Texas roadway.

Not only is the award-winning solution more environmentally friendly and more attractive, it is also more functional. The one-hundred-year flood is completely contained within the right-of-way without the roadway itself flooding. The roadway is also able to treat the "first flush," which is the first inch of stormwater runoff.

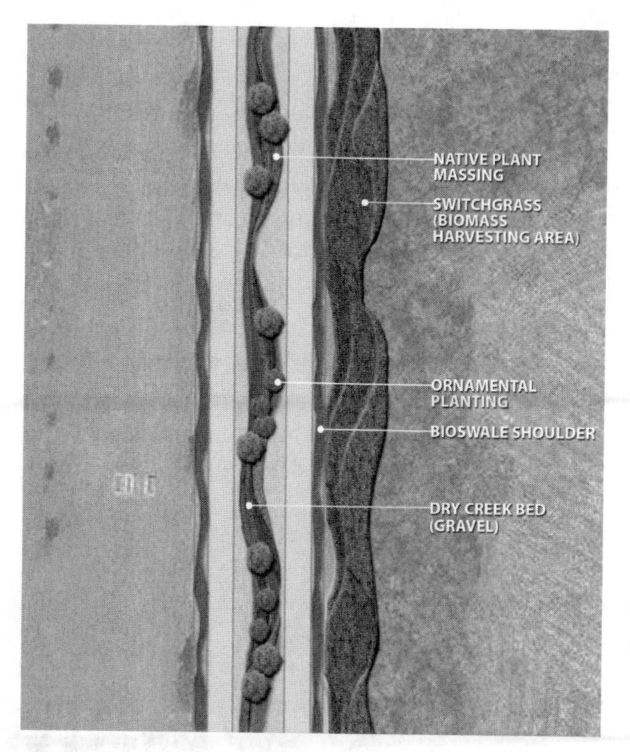

FIGURE 12-13 The landscape consists of strong, dynamic patterns that add visual interest to the corridor. Switchgrass planted in the same patterns along the side of Houston's Independence Parkway can be harvested for biofuel. Image courtesy of AECOM.

International Case Studies

The main reasons the United States is having difficulties planning for the next generation of transportation in this country is that we are so dependent upon the automobile, we have so many roads that are in disrepair, we are not meeting current transportation demands much less building what is needed for the future, and our public transportation system is limited.

We can learn a lot by studying what other countries are doing to plan for their future transportation infrastructure. China, with a population of 1.3 billion, is building a 53,000-mile national expressway system; India, with a population of one billion, is building a 10,000-mile national expressway system; and Europe, with a population of 450 million, is spending hundreds of billions of euros on a network of highways, bridges, tunnels, ports, and rail lines.

Japan has 1,250 miles of high-speed rail and is building about 190 more miles by 2020; China is planning to build more than 1,550 miles of high-speed rail by 2020; Europe has more than 2,500 miles of high-speed rail and is planning to build 560 more by 2020. The United States has about 190 miles and is building none (AASHTO, 2007). That needs to change.

International projects can provide examples of how other countries are addressing green roadway issues, and how their solutions may be incorporated into our next steps for creating green roadways.

Henderson Waves Bridge, Singapore

The Henderson Waves Bridge is located in Singapore, in the middle of the Southern Ridge, which is a 5.5-mile chain of hills from Mount Faber Park to Kent Ridge Park. The bridge illustrates an artistic, creative way to build a pedestrian bridge. The architect wanted to create a bridge that was more than just a walkway. The final design for the bridge, which was constructed in 2008, incorporates an undulating wave design that has made the bridge a visual landmark and popular tourist attraction.

The Henderson Waves Bridge specifically creates a pedestrian connection between Mount Faber Park and Telok Blangah Hill Park. It is the tallest pedestrian bridge in Singapore at 118 feet above Henderson Road. It is 899 feet long, with seven spans.

FIGURE 12-14 The Henderson Waves Bridge in Singapore is designed with seven undulating curved steel "ribs" that alternately rise over and under its deck. The unique waveform of the bridge gives it a distinct look. Image courtesy of Wikipedia Commons.

The bridge has the same basic structure as most pedestrian bridges, with beams supported by pillars. However, this bridge has a unique wave-shaped design. Each span has a curved "rib" as opposed to a normal flat bridge. The spans are a pattern of arches and catenaries (downward arches) that alternatively rise and dip along the bridge. Mathematical equations were used to create the simple, rhythmic form of the waves. The ribs of the bridge are constructed of steel to add structural integrity. The arches along the bridge form nooks along the walkway that are used as sitting areas and provide a warm, dry place to rest.

The deck of the bridge is made from yellow balau wood, which is native to the local environment. There are more than five thousand boards, and the degree of the boards changes every thirty-three feet to account for the curvature of the deck. The wooden decking material helps the bridge visually blend with the surrounding environment. At night, the deck and waves are lit with light-emitting diode lights that make the bridge a focal point of the area.

The Henderson Waves Bridge takes a different approach to creating a pedestrian bridge. It uses a unique design idea to appeal to the community and incorporate the existing landscape.

Figure 12-15 The lighting for the Henderson Waves Bridge helps enhance the visual character of the bridge at night. Image courtesy of Wikipedia Commons.

Figure 12-16 The curved ribs of Henderson Waves Bridge form alcoves that provide seating when needed for public events. Image courtesy of Wikipedia Commons.

Clem Jones Tunnel, Brisbane, Australia

The Clem Jones Tunnel (CLEM7) is part of the Brisbane, Australia's, M7 motorway, which links Brisbane's southern suburbs to the airport and fast-growing northern suburbs. The project is part of an overall strategy to improve the efficiency of Brisbane's road network. It is expected to reduce surface traffic, and in doing so, enable a series of urban enhancements in adjacent suburbs. Reducing surface traffic is one of the key concepts behind green roadways.

There are three distinctive man-made canopies above the entry and exit points of the CLEM7 tunnel. The canopies' dynamic, flowing forms are supposed to mimic the protection of Brisbane's expansive subtropical shade trees. Each canopy is supported by steel trusses covered by metallic copper-colored architectural panels, and the largest canopy consists of up of two thousand separate pieces.

The tunnel cost AUD$3.2 billion and is Brisbane's first privately financed inner-city toll road. The toll road is four miles long, and the tunnel is three miles long. Eighteen bridges allow for cross traffic over the tunnel. Construction began in September 2006, and the CLEM7 Community Open Day took place on February 28, 2010.

Toll collection is conducted by an electronic tolling system or by photographing license plates. One complaint about the tunnel is the high price of the tolls. The toll rates are among the most expensive in Australia, and they are scheduled to go up each year on the first of January for the next forty-five years. Tolling is becoming one of the most popular ways to pay for transportation projects in Australia, as it is in many other parts of the world.

The public-private partnership approach used for the project helped reduce the financial risk to the City of Brisbane. One concern, though, is that the number of vehicles using the tunnel has been significantly less than was originally projected. If the original estimates are not met in the future, it will be difficult to meet the economic projections for the project.

Zaragoza Bridge Pavilion, Zaragoza, Spain

The Zaragoza Bridge is a breathtaking pedestrian bridge built across the River Ebro in Zaragoza, Spain. The bridge, which was part of Spain's Expo 2008, also doubles as an exhibition space.

The Zaragoza Bridge is 918 feet long, and the piles that support the bridge are 225 feet tall, making them the tallest in Spain. It spans from one bank of the River Ebro to a small island in the river, and then continues on to the other riverbank. The bridge is built in four different sections, referred to as pods. Each pod is designed for a different function, and each has its own unique physical and spatial characteristics. Designing the bridge in pods allowed the bridge to be constructed with a modular approach that required smaller load-bearing members than would have been required otherwise. The pods also

FIGURE 12-17 The CLEM7 Tunnel at dusk, Brisbane, Australia. Image courtesy of RiverCity Motorway.

FIGURE 12-18 The CLEM7 Tunnel, built 197 feet below the Brisbane River, uses state-of-the-art monitoring to ensure the safety of passengers. Image courtesy of Shannon Walker.

FIGURE 12-19 The CLEM7 Tunnel was designed to alleviate traffic congestion in the city of Brisbane, bypassing twenty-four existing sets of traffic lights. Image courtesy of AECOM.

FIGURE 12-20 The Zaragoza Bridge Pavilion also doubles as a pedestrian bridge across the river Ebro in Zaragoza, Spain. Image courtesy of Wikipedia Commons.

serve to define the different use areas of the bridge so that each has its own unique identity.

The bridge is shaped like a diamond. After many studies, designers felt a diamond shape would be best for distributing the load of the bridge throughout the entire surface, thus allowing the the size of the support structures to be minimized. Additionally, with the walkway in place, the diamond shapes create a triangular area underneath that can serve as an area for utilities or other services.

The pods are strategically placed, with the longer span (607 feet) containing only one pod. From there, the other three pods are connected and continue the rest of the distance to the opposite riverbank. Each pod functions as a separate unit, and buffer zones help create a sense of separation between each pod. This helps create a more distinct transition as a visitor moves from one pod to the next. The longer pod and one of the smaller pods emphasize a more natural environment by including small triangular openings at the tops of the pods, with larger openings at the lower level. This allows visitors to have striking views of the river and the Expo. The other two pods are entirely enclosed.

The design of the exterior of the bridge pods looks like a pattern of shark scales. This pattern was chosen for several reasons: they are visual appealing, they can be wrapped around complex curved shapes, and they can be constructed relatively inexpensively. Although the use of the shark-scale pattern is consistent

FIGURE 12-21 Interlocking of the pods in the Zaragoza Bridge Pavilion allows visitors to move from pod to pod through small in-between spaces that act as buffer zones. Image courtesy of Wikipedia Commons.

across the entire structure, the material is split longitudinally. The upper half of the exterior is made from glass-reinforced concrete and the lower half is made of metal plates. There are various shades, ranging from white to black, throughout the bridge.

Atlantic Road Bridge, Norway

The Atlantic Road (Atlanterhavsveien) in Norway connects the mainland Romsdal peninsula to the island of Averøya. The road is five miles long and connects a series of small islands with landfills and eight bridges. Construction on the road started in August 1983, and the road opened on July 7, 1989. In 2005 the road was voted Norwegian Construction of the Century.

The road twists and turns as it weaves from one tiny island to the next. The tallest bridge along the road is the Storseisundet Bridge, which has a dramatic, curving shape that seems to soar and twist over the water. The Storseisundet is the tallest and the longest (at 853 feet) of the eight bridges. The bridge seems to be more a piece of artwork than it does an engineered structure.

The road is a popular tourist attraction in large part because of the dramatic views of the sea. It is the second most-visited scenic road in the country and is part of a twenty-two-mile-long National Tourist Route. Driving the Atlantic Road during the fall is especially popular because the storms in the area result in big waves that make the drive even more dramatic.

There are four scenic viewing areas along the road, and each provides panoramic views of the islands, villages, and surrounding sea. Despite the lack of shoulders, fishing from the bridges is a popular recreational activity. Fishing also occurs from the viewing areas.

FIGURE 12-22
Norway's Storseisundet Bridge, located along "The Road to Nowhere," is the longest of the eight bridges that make up the Atlantic Road. Image courtesy of Wikipedia Commons.

FIGURE 12-23 The Storseisundet Bridge is built so that, as you approach certain angles, it looks as if you may plummet off its seventy-five-foot height. Image courtesy of Wikipedia Commons.

City of Saskatoon (Canada) Green Streets Program

Saskatoon is the largest city in the Canadian province of Saskatchewan. In 2009, the city developed a Green Streets program to promote innovative roadway engineering that included the use of recycled materials. The city had a large stockpile of asphalt and concrete rubble materials, so finding a way to recycle this material made sense. In 2006, they initiated research to determine the best way to use the materials.

The city is now producing a variety of crushed concrete materials and asphalt aggregate materials that are superior to locally available aggregates. The result is roadways that are superior, are less expensive, and recycling saves money, makes good use of waste material, and keeps the materials out of landfills.

The city has estimated that the Green Streets program reduces costs by around 45 percent as compared to traditional ways of building roads using virgin-sourced aggregates. There is also a significant reduction in the energy and resources needed to create the roads, resulting in a smaller carbon footprint.

In early 2009, the City of Saskatoon constructed two streets under the Green Streets program. The project required approximately 1,984 tons of recycled concrete and asphalt and 595 tons of an asphalt material called Duraclime. Duraclime is manufactured by Lafarge, a leading developer and supplier of asphalt and concrete products in the United States and Canada. Duraclime is manufactured using recycled asphalt and a warm-mix asphalt technology that minimizes emissions, fuel consumption, smoke, and odors during the manufacturing process because of the lower temperatures used.

The concrete used for curbs and sidewalks meets LEED specifications, and the aggregates used for the project's base materials are 100 percent recycled concrete for the subbase and 100 percent recycled asphalt for the base.

Summary

In the United States, we are an automobile-oriented society, and that isn't likely to change anytime soon.

Figure 12-24 Colorful autumn leaves grace this winding road through a rural area of Saskatoon. Image acquired from depositphotos.com.

FIGURE 12-25 An aerial view of the main road (with no traffic) in Zaragoza, Spain. Image acquired from depositphotos.com.

Much of our existing infrastructure is badly in need of repair, and we will need new roads to accommodate growth. Traffic congestion continues to get worse every year as we put more and more cars on existing roads. As we build new roads, and rebuild existing roads, we need to think of roads as an integral part of a community or landscape. We can build roads that are safe and effective, but that also look good, are respectful of the environment, and enhance community character.

Discussion of green roadways is not just about the physical aspects of highways and roadways. It is about how people get from one place to another, whether that be in their cars, on a bus or train, or on a bicycle or other nonmotorized vehicle. To make our road system work, we need to take a more integrated approach to designing and building roads, but we also need to get people out of their cars. Creating walkable communities where people can walk to work or walk to a neighborhood park to play will greatly reduce demand on our roads. The subdivisions of the 1950s and '60s, which were built around the automobile, no longer work. It is time to rethink the concept

FIGURE 12-26 This French village is nestled in a large farm area, as seen from an aerial view of the countryside. Image acquired from depositphotos.com.

Figure 12-27 In Strasbourg, Germany, the right-of-way for the tramway is grass, helping give the city a greener visual appearance. Image acquired from depositphotos.com.

of subdivisions just like we are rethinking the idea of multiuse development in urban areas.

By taking an integrated, holistic approach to transportation, we can provide options so that people are less dependent upon their cars. And when we do build roads, as this book has shown, it is possible to make sure they are environmentally friendly, safe, efficient, cost-effective, durable, long-lasting, and fit their environment. It is all about designing a transportation infrastructure that improves quality of life for current and future generations.

APPENDIX 1: RESOURCE CHARACTERISTICS

To construct this chart, assumptions were made. Speeds, traffic, amount of heavy vehicles, and level of service, along with many other factors, must be taken into consideration. Each roadway is different, with unique characteristics and different challenges and the standards might not fit with every road. This chart is only to be used to gain a better understanding of the different types of roadways.

	Local/Collector Road				Arterials				Freeways			
	Urban		Rural		Urban		Rural		Urban		Rural	
	Desired	Required	Desired	Required	Desired	Required	Desired	Required	Desired	Required	Desired	Required
Lane width (feet)	12	9	12	9	12	10	12	11	12	12	12	12
Lane slope (%)	2	<6	3	<6	2	2	2	2	2	2	2	2
Shoulder width (feet)	4–8	2	4	2	4–8	4	4–8	4	4–8 left; 12 right	4 left; 10 right	4–8 left; 12 right	4 left; 10 right
Shoulder slope (%)	4		4		4		4		4		4	
# of lanes	2		2		Varies		Varies		4–6		4–6	
Curb & gutter	as needed		almost never		as needed		rarely		Only in special cases		Only in special cases	
Clear zone (feet)	10	10	10	10	30	Varies	30	Varies	30	Varies	30	Varies
Horizontal curve (radius, feet)	Based on speed		Based on speed		Based on speed		Based on speed		Based on speed		Based on speed	
Parking	7' lanes	—	None				None		None	None	None	None
Design speeds (mph)	20–30	—	30–50	—	30–60		40–75		70	>50	70	>50
Sidewalk width (feet)	4	4			5	4	5	4				
Bike lane width (feet)	4	4			4	4	4	4				
Longitudinal grade (%)	<5	<15	<5	<17	<5	<11	<3	<8	<6	<3	<6	<3
Median width (feet)	Desired, but varies greatly		None normally		Varies greatly		Varies greatly		10	Varies greatly	15–30	Varies greatly

SELECTED RESOURCES

"11 Most Endangered Historic Places: Merritt Parkway." National Trust for Historic Preservation. www.preservationnation.org/travel-and-sites/sites/northeast-region/merritt-parkway.html.

1000 Friends of Florida. 2001. *Wildlife Habitat Planning Strategies, Design Features, and Best Management Practices for Florida Communities and Landowners. Florida's Wildlife Legacy Initiative*. Dec. 11. 1000fof.org/Panhandle/Documents/Wildlife/pennington12-11-071.ppt.pdf.

AASHTO. 2010. *The Road to Livability*. Washington, D.C.

AASHTO. 2011. *A Policy on Geometric Design of Highways and Streets*, 6th ed.

"Alligator Alley." Wikipedia. en.wikipedia.org/wiki/Alligator_Alley.

American Association of State Highway and Transportation Officials' (AASHTO). 2007. *Transportation—Invest in Our Future*. www.transportation1.org/tif5report/intro.html.

American Society of Civil Engineers. 2009. Roads–Report Card for America's Infrastructure. www.infrastructurereportcard.org/fact-sheet/roads.

Ariniello, A., and B. Przybyl. 2010. "Roundabouts and Sustainable Design." *Proc. of Green Streets and Highways 2010*. American Society of Civil Engineers. ascelibrary.org/proceedings/resource/2/ascecp/389/41148/8_1?isAuthorized=no.

"The Atlantic Road." *Your Travel Guide to Norway*. www.bestnorwegian.com/atlantic_road.html.

Bryce, J. M. 2008. "Exploring Green Highways." *Standardization News*. Sept.-Oct. www.astm.org/SNEWS/SO_2008/bryce_so08.html.

California Department of Transportation. *Highway Design Manual*. www.dot.ca.gov/hq/oppd/hdm/hdmtoc.htm.

Cassidy, W. B. 2010. "LaHood Calls for 'Creative' Infrastructure Funding." *The Journal of Commerce Online*. June 10. www.joc.com/government-regulation/lahood-calls-%E2%80%98creative%E2%80%99-infrastructure-funding.

CH2M Hill with EDAW. 2002. A Guide to Best Practices for Achieving Context-Sensitive Solutions. National Cooperative Highway Research Program Report 480. Transportation Research Board.

Chew, R. 2009. "'Green Highways': Michigan's Eco-Friendly Road Concept." *Time Magazine*. Oct. 10. www.time.com/time/health/article/0,8599,1928997,00.html.

Cho, A., and T. Newcomb. 2010. Oregon Project Is Case Study for "Green" Road Standards. July 22. northwest.construction.com/northwest_construction_news/2010/0722_OregonProject.asp.

"Clem Jones Tunnel." Wikipedia. en.wikipedia.org/wiki/Clem_Jones_Tunnel.

"Competition To Develop Innovative Design Solutions." ARC International Wildlife Crossing Infrastructure Design. www.arc-competition.com/files/ARC_Brief.pdf.

Complete Streets. www.completestreets.org.

Council on Environmental Quality. 2007. *Having Your Voice Heard*. Executive Office of the President. Dec. ceq.hss.doe.gov/nepa/Citizens_Guide_Dec07.pdf.

Cramer, P., and J. A. Bissonette. 2009. "Transportation Ecology and Wildlife Passages—The State of the Practice and Science of Making Roads Better for Wildlife." *TR News*. May–June. Number 262. onlinepubs.trb.org/onlinepubs/trnews/trnews262.pdf.

Dahan, O., and A. Goykhman. 2009. "The Importance of Green Roads." GreenBiz.com. Aug. 22. www.greenbiz.com/blog/2009/08/26/importance-green-roads.

"The Delaware Avenue Extension, Philadelphia's Green

Road, Philadelphia Sustainability Awards." Philadelphia Sustainability Awards. www.philadelphiasustainability awards.org/nominees/volmer.

Delaware Department of Transportation, Division of Planning and Policy. 2000. *Traffic Calming Design Manual—Final Regulations*. Sept. 1. www.deldot.gov/information/pubs_forms/manuals/traffic_calming/pdf/deldotfinal.pdf.

Design Workshop. 2006. I-15 Corridor Study. Nevada Department of Transportation.

Doster, A., and K. Sheppard. 2009. "The Future of Transit—Public Transportation Needs Massive Investment. Will the Obama Administration Step Up?" *In These Times*. Feb. 23. www.inthesetimes.com/article/4262/the_future_of_transit/.

The Economist. 2011. "America's Transport Infrastructure: Life in the Slow Lane." April 28. www.economist.com/node/18620944.

Ehl, L. 2009. Obama Calls for "More Creative" Ways to Pay for Infrastructure. Washington State Department of Transportation. Federal Transportation Issues. wsdot federalfunding.blogspot.com/2009_11_01_archive.html.

FHWA. n.d. FHWA and Context Sensitive Solutions. www.fhwa.dot.gov/csd.

FHWA. n.d. "Interstate 15 / Blue Diamond Interchange." Context Sensitive Solutions.org. FWHA. www.context sensitivesolutions.org/content/case_studies/interstate_15_blue_diamond_in/.

FHWA. n.d. U.S. Department of Transportation. *Flexibility in Highway Design*. www.fhwa.dot.gov/environment/flexibility/ch01.cfm.

FHWA. n.d. U.S. Department of Transportation. *Planning for Transportation in Rural Areas*. www.fhwa.dot.gov/Planning/rural/planningfortrans/2ourrts.html.

FHWA. n.d. Water, Wetlands, and Wildlife. Environmental Review Toolkit. Stormwater Best Management Practices in an Ultra-Urban Setting: Selection and Monitoring. environment.fhwa.dot.gov/ecosystems/ultraurb/uubmp3p1.asp.

FHWA. 1996. U.S. Department of Transportation. *Public Involvement Techniques for Transportation Decision-making*. FHWA-PD-96-031. www.fhwa.dot.gov/reports/pittd/cover.htm.

FHWA, 1997. Office of Planning, Environment, and Realty. *Flexibility in Highway Design*. www.fhwa.dot.gov/environment/flexibility/.

FHWA. 1999. U.S. Department of Transportation. Building Roads in Sync with Community Values. H. Peaks and S. Hayes. *Public Roads*. Vol. 62, No. 5, Mar.–Apr. 1999. www.fhwa.dot.gov/publications/publicroads/99marapr/flexdsgn.cfm.

FHWA. 2002. U.S. Department of Transportation. A Hallmark of Context-Sensitive Design. S. Moler. *Public Roads*. Vol. 6, No. 6. May–June. www.fhwa.dot.gov/publications/publicroads/02may/02.cfm.

FHWA. 2003. Environmental Review Toolkit. Streamlining and Stewardship. Preserving a Sense of Place: Kentucky's Paris Pike Project. US Department of Transportation. Oct. www.environment.fhwa.dot.gov/strmlng/newsletters/oct03nl.asp.

FHWA. 2004. U.S. Department of Transportation. A Tale of Two Canyons. S. Moler. *Public Roads*. Vol. 67, No. 5. Mar.–Apr. www.fhwa.dot.gov/publications/publicroads/04mar/05.cfm.

FHWA. 2007. Federal Transit Administration. *The Transportation Planning Process: Key Issues. A Briefing Book for Transportation Decisionmakers, Officials, and Staff*. Transportation Planning Capacity Building Program, Sept. FHWA-HEP-07-039. www.planning.dot.gov/documents/briefingbook/bbook.htm.

FHWA. 2008. "Maryland's Intercounty Connector: Using Environmental Stewardship to Redefine Project Management." *Successes in Stewardship*. environment.fhwa.dot.gov/strmlng/newsletters/feb08nl.asp.

FHWA. 2009. U.S. Department of Transportation. Building Lightly on the Land. A. Armstrong, H. G. Armstrong, and K. J. Smith. *Public Roads*. March/April. Vol. 72, No. 5. www.tfhrc.gov/pubrds/09mar/02.htm.

Fisher, M. 2008. "Green Highways: Research Targets Environmentally Friendly Asphalts." News from the University of Wisconsin-Madison. May 27. www.news.wisc.edu/15277.

Foth, M., D. Guenther, R. Haichert, and C. Berthelot. 2010. City of Saskatoon's Green Streets Program—A Case Study for the Implementation of Sustainable Roadway Rehabilitation with the Reuse of Concrete and Asphalt Rubble Materials. pp. 337–348. *Green Streets and Highways, 2010*. American Society of Civil Engineers. cedb.asce.org/cgi/WWWdisplay.cgi?271754.

Fountain, H. 2009. "Concrete Is Remixed With Environment in Mind." *New York Times*. Mar. 30. www.nytimes.com/2009/03/31/science/earth/31conc.html?_r=1.

Franklin Regional Council of Governments, Franklin Regional Planning Board. 2002. *Design Alternatives for Rural Roads*. Greenfield, MA: Franklin Regional Council of Governments.

Freemark, Y. 2010. "The Future of Streetlights: 6 Brilliant New Concepts." Infrastructurist: America Under Construction. Apr. 28. www.infrastructurist.com/2009/04/28/the-future-of-streetlights-6-brilliant-new-concepts/.

Gelinas, N. 2007. "Lessons of Boston's Big Dig." *City Journal*. Autumn. www.city-journal.org/html/17_4_big_dig.html.

Gray, T. B. 1999. Chapter Two: The History of Highway Aesthetics. *The Aesthetic Condition of the Urban Freeway*. www.mindspring.com/~tbgray/prch2.htm.

"Green Planning." 2010. *Environmental Business Journal*, Vol. XXIII, No. 04. *Environmental Business International*. Apr. www.ebionline.org/downloadenvironmental-business-journal/file/view/30/6.

"The Henderson Waves Bridge–Singapore." *The Best Travel Destinations*. thebesttraveldestinations.com/the-henderson-waves-bridge-singapore/.

Highland Bridge. Carter & Burgess, Inc. www.ite.org/activeliving/files/C-2-C_ppa016.pdf.

Kelly, M. "Eco-friendly Road Construction." The University of Texas at Austin. herbie.ischool.utexas.edu/~lkh577/templates/site/research/articles/eco.html.

Lardner/Klein Landscape Architects et al. 2008. "US Route 4 Corridor Management Plan, Hartford, Vermont." Vermont Department of Transportation, Sept. www.aot.state.vt.us/planning/Documents/Planning/U.S.%204_Final%20Report.pdf.

Leighton Contractors Pty Limited. 2009. "CLEM Jones Tunnel." Brisbane City Council, 26 June. www.investmentfacilitymanagement.com.au/downloads/CLEM7%20Data%20Sheet_V1.0.pdf.

Lendman, S. "Predicting Worse Ahead From the US Economic Crisis," Sept. 9, 2009. rense.com/general87/lindd.htm

Lynch, J. 2010. "The Greenway Way." *Issaquah & Sammamish Reporter*. July 8. www.pnwlocalnews.com/east_king/iss/news/98038279.html.

Marek, M. A. 2007. *Landscape and Aesthetics Design Manual*. Texas Department of Transportation. onlinemanuals.txdot.gov/txdotmanuals/lad/manual_notice.htm.

Maryland Department of Transportation. 1998. Thinking Beyond the Pavement Conference. Context Sensitive Solutions and Complete Streets. sha.md.gov/Index.aspx?PageId=332.

McGowen, P. T., and M. P. Huijser. 2009. "Best Practices for Reducing Wildlife–Vehicle Collisions." *TR News*. May–June. www.trb.org/main/blurbs/161826.aspx.

"Merritt Parkway." Wikipedia. en.wikipedia.org/wiki/Merritt_Parkway.

Merritt Parkway Working Group. *Merritt Parkway Guidelines for General Maintenance and Transportation Improvements*. Newington, CT: Connecticut Dept. of Transportation, 1994.

Michigan DOT. 2010. *MDOT Opens State-of-the-art Pedestrian Bridge, Reunites Detroit's Mexicantown Community*. B. V. Peek. May 05. michigan.gov/mdot/0,1607,7-151-9620_11057-236536–RSS,00.htm.

Mountains to Sound Greenway. mtsgreenway.org.

National Cooperative Highway Research Program. 2007. *Guidelines for the Selection of Snow and Ice Control Materials to Mitigate Environmental Impacts*. Report 577. Transportation Research Board. onlinepubs.trb.org/onlinepubs/nchrp/nchrp_rpt_577.pdf.

National Cooperative Highway Research Program. 2007. *Roundabouts in the United States*. Report 572. Transportation Research Board.

National Park Service. 1991. Developing Sustainable Mountain Trails. Rocky Mountain Region.

Natural Resources Defense Council (NRDC). 2000. *Paving Paradise*. www.nrdc.org/cities/smartgrowth/rpave.asp.

Nevada DOT. *n.d. Henderson Interstate 215/515 Interchange*. Context Sensitive Solutions. FHWA. contextsensitivesolutions.org/content/case_studies/henderson_interstate_215_515_in/.

Nevada DOT. 2002. *Pattern and Palette of Place: A Landscape and Aesthetics Master Plan for the Nevada State Highway System*. July 3. www.nevadadot.com/uploadedFiles/MasterPlan-July3.pdf.

New York State. Department of Transportation. *Context Sensitive Solutions (CSS) Engineering Instruction EI 01-020*. www.dot.ny.gov/divisions/engineering/design/dqab/css/repository/enginst.pdf?nd=nysdot.

North Carolina Department of Transportation. 2008. North Carolina Turnpike Authority. *Financing, Building, and Operating Tomorrow's Roads Today*. Dec. 10. www.ncdot.gov/turnpike/download/FINAL_JLTOC_AnnualPresentationV3NoVideo.pdf.

Oberstar, J. L., J. L. Mica, P. A. DeFazio, and J. J. Duncan, Jr. 2009. *The Surface Transportation Authorization Act of 2009—A Blueprint for Investment and Reform*. U.S. House of Representatives. Committee on Transportation and Infrastructure. June 18. cmaanet.org/files/shared/Oberstar_Bill_Summary.pdf.

"Oregon Solar Highway." 2012. www.oregon.gov/ODOT/HWY/OIPP/inn_solarhighway.shtml.

Paniati, J. 2009. "Transportation Infrastructure Financing Opportunities and Challenges." Proc. Fifth Annual Public Private Partnerships USA Summit, Mar. 12. Federal Highway Administration. www.fhwa.dot.gov/pressroom/re090312.htm.

"Paris Pike, Kentucky." National Trust for Historic Preservation. www.preservationnation.org/resources/case-studies/transportation/parispike-ky-4f-case-study.pdf.

Perez, K. 2010. "Roofs to Roads: Recycled Shingles Make Paving Projects Greener." Boulder Daily Camera. June 24. www.dailycamera.com/boulder-county-news/ci_15371770.

"Rainwater Gardens." Maplewood, Minnesota. www.ci.maplewood.mn.us/index.aspx?NID=456.

The Road to Better Transportation Projects: Public Involvement and the NEPA Process. The Sierra Club and Natural Resources Defense Council. www.sierraclub.org/sprawl/nepa/sprawl_report.pdf.

"S. Korea Unveils 'Recharging Road' for Eco-friendly Buses." 2010. Phys.org. Mar. 09. www.physorg.com/news187331386.html.

Schoen, J. W. 2007. "U.S. Highways Badly in Need of Repair." MSNBC, Aug. 3. www.msnbc.msn.com/id/20095291/.

Seattle Housing Authority. 2006. *Green Home Case Study*. City of Seattle Department of Planning and Development. Jan. www.seattle.gov/DPD/cms/groups/pan/@pan/@sustainablebldig/documents/web_informational/dpds_007254.pdf.

Seattle Housing Authority. 2007. *High Point Redevelopment Project*. www.brunerfoundation.org/rba/pdfs/2007/high_point.pdf.

Sipes, J. L. 2001. "New Respect for the Land Drives Highway Design." *Seattle Daily Journal*. Apr. 19. www.djc.com/news/enviro/11121025.html.

Sipes, J. L. 2003. "Bridges That 'Fit': Designing Bridges That Fit Their Surroundings Both Environmentally and Aesthetically." *Landscape Architecture* 93 (2): 46.

Sipes, James L. 2006. Principles of Smart Growth. Unpublished manuscript.

Sipes, J. L. 2008. Road to Ruin: The Changing Landscape of Transportation Planning, unpublished manuscript.

Sipes, J. L., and D. Fasser. 2001. "Context Sensitive Transportation Planning and Design." Proc. of 2001 American Society of Landscape Architects Annual Meeting Proceedings. 100-03.

Sipes, J. L., and J. Neff. 2001. "Fencing, Wildlife Crossings, and Roads: Separating Animals and Vehicles." *Landscape Architecture*. June. Vol. 91, No. 6: 24–27.

Sipes, J. L., and R. Blakemore. 2007. "Aesthetics in the Landscape—How Nevada and Other States Are Integrating Aesthetics into Transportation Projects." *TR News*. Jan.–Feb. onlinepubs.trb.org/onlinepubs/trnews/trnews248aesthetics.pdf.

Skinner, S. 2010. "Road Sustainability—Environmentally Friendly Machines and Processes for Road Building." *Construction Europe*. Feb. 15. www.khl.com/magazines/construction-europe/detail/item51430/Road-sustainability—environmentally-friendly-machines-and-processes-for-road-building/.

Snelling, S. A. 2010. "Towards Green Bridges." *Proc. Transportation Research Board 89th Annual Meeting*, Transportation Research Board. trid.trb.org/view.aspx?id=909635.

Sousa, L. R., and J. Rosales. 2010. "Contextually Complete Streets." *Proc. ASCE Green Streets and Highways Conference*. American Society of Civil Engineers. content.asce.org/files/pdf/Sousa.pdf.

Squatriglia, C. 2010. "Finland Proposes World's First 'Green Highway.'" *Wired*. Aug. 20. www.wired.com/autopia/2010/08/finland-green-highway/.

Texas Department of Transportation. Landscape Design Section, Design Division. 2001. *Aesthetics in Transportation Design: Color and Texture Alternatives*. Jan. 22. Web. www.dot.state.tx.us/ insdtdot.orgchart/des/landscape.

Transportation Research Board, National Cooperative Highway Research Program. 1993. Recommended Procedures for the Safety Performance Evaluation of Highway Features. Report 350. onlinepubs.trb.org/onlinepubs/nchrp/nchrp_rpt_350-a.pdf.

Transportation Research Board. Access Management Manual. 2003. www.trb.org/Main/Blurbs/152653.aspx.

University of North Carolina. 2010. "Hundreds Celebrate First Walk across Bagley Pedestrian Bridge." *Wayne County News*. May 28. www.metromode.com/waynecounty/innovationnews/three0014.aspx?referrerID=a5dd6703-2797-4d9e-ba52-a645803bdca3.

Urban Land Institute and Ernst & Young, LLP. 2007. *Infrastructure 2007: A Global Perspective*. Washington, D.C.

U.S. Energy Information Agency. 2008. Short-Term Energy Outlook. Feb.

U.S. General Accounting Office. 2012. "Traffic Congestion—Road Pricing Can Help Reduce Congestion, but Equity Concerns May Grow." www.gao.gov/assets/590/587833.pdf.

U.S. Government Accountability Office. 2000. Determining Performance and Accountability Challenges and High Risks. www.gao.gov/assets/210/200448.pdf.

"Vancouver Land Bridge: Project Team." 2008. *Daily Journal of Commerce*. Apr. 25. djcoregon.com/news/2008/04/25/vancouver-land-bridge-project-team.

Venner, M., and A. Santalucia. 2010. *Environmental Corridor Management*, NCHRP 25-25/63. AASHTO, Standing Committee on Environment. onlinepubs.trb.org/onlinepubs/nchrp/docs/NCHRP25-25(63)_FR.pdf.

Washington State DOT. 2007. Best Practices in Corridor Planning. K. Lindquist, June 11. www.wsdot.wa.gov/NR/rdonlyres/8A3F36D8-FEF0-4E38-8547-CD07C7F07D7A/0/CorridorStudies1.pdf.

Washington State DOT. 2010. *Context Sensitive Design Can Reflect Community Interests*. Federal Transportation Issues., 17 June 2010. wsdotfederalfunding.blogspot.com/2010/06/context-sensitive-design-can-reflect.html.

Wenzel, D. 2009. "Greening America's Roadways." *Planetizen*. July 06. www.planetizen.com/node/39545.

Witkin, J. 2011. "Pod Cars Start to Gain Traction in Some Cities." *New York Times*. Sept. 20. wheels.blogs.nytimes.com/2010/09/20/pod-cars-start-to-gain-traction-in-some-cities/.

World Economic Forum. 2011. Sustainable Credit Report 2011. www.weforum.org/reports/sustainable-credit-report-2011.

Zaha Hadid Architects. Zaragoza Bridge Pavilion. www.zaha-hadid.com/design/zaragoza-bridge-pavilion/.

Zaragoza Bridge Pavilion. 2008. Expoagua Zaragoza 2008. www.zaha-hadid.com/wp-content/files_mf/zaragoza.pdf.

INDEX

Note: Figures/photos/illustrations are indicated by an "f".

Practicar

Eureka Math®

1.er grado
Módulos 1–3

Publicado por Great Minds®.

Impreso en los EE. UU.
Este libro puede comprarse en la editorial en eureka-math.org.
10 9 8 7 6 5 4 3 2

v1.0 PAH

ISBN 978-1-64054-872-5

G1-SPA-M1-M3-P-05.2019

Aprender • Practicar • Triunfar

Los materiales del estudiante de *Eureka Math*® para *Una historia de unidades*™ (K–5) están disponibles en la trilogía *Aprender, Practicar, Triunfar*. Esta serie apoya la diferenciación y la recuperación y, al mismo tiempo, permite la accesibilidad y la organización de los materiales del estudiante. Los educadores descubrirán que la trilogía *Aprender, Practicar y Triunfar* también ofrece recursos consistentes con la Respuesta a la intervención (RTI, por sus siglas en inglés), las prácticas complementarias y el aprendizaje durante el verano que, por ende, son de mayor efectividad.

Aprender

Aprender de *Eureka Math* constituye un material complementario en clase para el estudiante, a través del cual pueden mostrar su razonamiento, compartir lo que saben y observar cómo adquieren conocimientos día a día. *Aprender* reúne el trabajo en clase—la Puesta en práctica, los Boletos de salida, los Grupos de problemas, las plantillas—en un volumen de fácil consulta y al alcance del usuario.

Practicar

Cada lección de *Eureka Math* comienza con una serie de actividades de fluidez que promueven la energía y el entusiasmo, incluyendo aquellas que se encuentran en *Practicar* de *Eureka Math*. Los estudiantes con fluidez en las operaciones matemáticas pueden dominar más material, con mayor profundidad. En *Practicar*, los estudiantes adquieren competencia en las nuevas capacidades adquiridas y refuerzan el conocimiento previo a modo de preparación para la próxima lección.

En conjunto, *Aprender* y *Practicar* ofrecen todo el material impreso que los estudiantes utilizarán para su formación básica en matemáticas.

Triunfar

Triunfar de *Eureka Math* permite a los estudiantes trabajar individualmente para adquirir el dominio. Estos grupos de problemas complementarios están alineados con la enseñanza en clase, lección por lección, lo que hace que sean una herramienta ideal como tarea o práctica suplementaria. Con cada grupo de problemas se ofrece una Ayuda para la tarea, que consiste en un conjunto de problemas resueltos que muestran, a modo de ejemplo, cómo resolver problemas similares.

Los maestros y los tutores pueden recurrir a los libros de *Triunfar* de grados anteriores como instrumentos acordes con el currículo para solventar las deficiencias en el conocimiento básico. Los estudiantes avanzarán y progresarán con mayor rapidez gracias a la conexión que permiten hacer los modelos ya conocidos con el contenido del grado escolar actual del estudiante.

Estudiantes, familias y educadores:

Gracias por formar parte de la comunidad de *Eureka Math*®, donde celebramos la dicha, el asombro y la emoción que producen las matemáticas. Una de las formas más evidentes de demostrar nuestro entusiasmo son las actividades de fluidez que ofrece Practicar de *Eureka Math*.

¿En qué consiste la fluidez en matemáticas?

Es natural asociar *fluidez* con la disciplina de lengua y literatura, donde se refiere a hablar y escribir con facilidad. Desde prekínder hasta 5.º grado, el currículo de *Eureka Math* ofrece diversas oportunidades, día a día, de consolidar la fluidez *en matemáticas*. Cada una de ellas está diseñada con el mismo concepto—aumentar la habilidad de todos los estudiantes de usar las matemáticas *con facilidad*—. El ritmo de las actividades de fluidez suele ser rápido y energético, celebrando el avance y concentrándose en el reconocimiento de patrones y asociaciones en el material. Estas actividades no tienen como objetivo dar calificaciones.

Las actividades de fluidez de *Eureka Math* brindan una práctica diferenciada a través de diversos formatos—algunas se realizan en forma oral, otras emplean materiales didácticos, otras utilizan una pizarra personal y otras incluso usan una guía de estudio y el formato de papel y lápiz—. *Practicar* de *Eureka Math* brinda a cada estudiante ejercicios de fluidez impresos correspondientes a su grado.

¿Qué es un Sprint?

Muchas de las actividades de fluidez impresas utilizan el formato denominado Sprint. Estos ejercicios desarrollan la velocidad y la exactitud en las destrezas que ya se han adquirido. Los Sprints, que se utilizan cuando los estudiantes ya están alcanzando un nivel de dominio óptimo, aprovechan el ritmo para provocar una pequeña descarga de adrenalina que aumenta la memoria y la retención. El diseño deliberado de los Sprints los hace diferenciados por naturaleza; los problemas van de sencillos a complejos, donde el primer cuadrante de los problemas es el más sencillo y la complejidad aumenta en los cuadrantes subsiguientes. Además, los patrones intencionales en la secuencia de los problemas obligan a los estudiantes a aplicar un razonamiento de nivel superior.

El formato sugerido para trabajar con un Sprint requiere que el estudiante realice dos Sprints consecutivos (identificados como A y B) para la misma destreza, en el lapso cronometrado de un minuto cada uno. Los estudiantes hacen una pausa entre los Sprints para expresar los patrones que identificaron al trabajar en el primer Sprint. El reconocimiento de patrones suele mejorar naturalmente el rendimiento en el segundo Sprint.

También es posible llevar a cabo los Sprint sin cronometrar el tiempo. Se recomienda especialmente no utilizar el cronometraje cuando los estudiantes aún están adquiriendo confianza en el nivel de complejidad del primer cuadrante de los problemas. Una vez que todos los estudiantes se encuentran preparados para llevar a cabo los Sprint con éxito, suele resultar estimulante y positivo comenzar a trabajar para mejorar la velocidad y la exactitud, aprovechando la energía que produce el uso del cronómetro.

¿Dónde puedo encontrar otras actividades de fluidez?

La *Edición del maestro* de *Eureka Math* guía a los educadores en el uso de las actividades de fluidez de cada lección, incluso aquellas que no requieren material impreso. Además, a través de *Eureka Digital Suite* se puede acceder a las actividades de fluidez de todos los grados, y es posible hacer una búsqueda por estándar o lección.

¡Les deseo un año colmado de momentos "¡ajá!"!

Jill Diniz

Jill Diniz
Jill Diniz Directora de matemáticas
Great Minds

Contenido

Módulo 1

Módulo 2

Módulo 3

1.er grado
Módulo 1

A

Respuestas

Nombre ______________________________ Fecha ______________

*Escribe el número de puntos. ¡Encuentra 1 o 2 grupos que te hagan encontrar el número total de puntos más fácil!

1.	●●		16.	●●●●● ●●●●	
2.	●●●		17.	●●●●● ●●●	
3.	●●●●		18.	●●●●● ●●●●●	
4.	●●●		19.	●●●●● ●●	
5.	●		20.	●●●●● ●	
6.	●●●●		21.	●●●●● ●●●●	
7.	●●●●●		22.	●●●●● ●●●●●	
8.	●●●●		23.	●●●● ●●●●●	
9.	●●●●● ●		24.	●●●●● ●●●	
10.	●●●●● ●●		25.	●●● ●● ●●●●●	
11.	●●●●●		26.	●●●●● ●●	
12.	●●●●		27.	●●● ●● ●● ●●●	
13.	●●●●● ●		28.	●● ●● ● ●● ●●	
14.	●●●●● ●●●		29.	●● ●●● ● ●●	
15.	●●●●● ●●		30.	●● ● ●● ●● ● ●●	

B

Respuestas correctas:

Nombre ______________________ Fecha ____________

*Escribe el número de puntos. ¡Encuentra 1 o 2 grupos que te hagan encontrar el número total de puntos ¡más fácil!

1.	●		16.	●●●●● ●●●	
2.	●●		17.	●●●●● ●●●●	
3.	●		18.	●●●●● ●●	
4.	●●●●		19.	●●●●● ●●●	
5.	●●●		20.	●●●●● ●●●●●	
6.	●●●●●		21.	●●●●● ●●●●	
7.	●●●●		22.	●●●●● ●●●●●	
8.	●●●●●		23.	● ●●●● ●●●●●	
9.	●●●●● ●●		24.	●●●●● ●●●●●	
10.	●●●●● ●		25.	●● ●●●●●	
11.	●●●●● ●●●		26.	●●● ● ●● ●●	
12.	●●●●● ●		27.	●● ●●● ●●● ●●	
13.	●●●●●		28.	●● ● ● ● ●● ●●	
14.	●●●●● ●●		29.	●● ●● ● ●● ●●	
15.	●●●●● ●		30.	●● ● ●●●● ●●	

Nombre ______________________________ Fecha __________

¡Carrera de vínculo numérico!

Haz tantos como puedas en 90 segundos. Escribe los vínculos numéricos que terminaste aquí:

1.
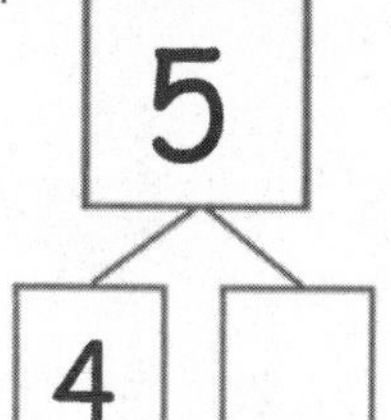

2.
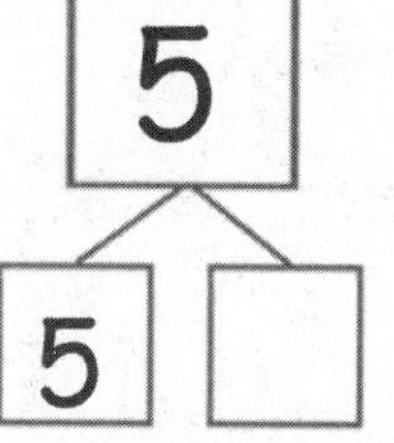

3.
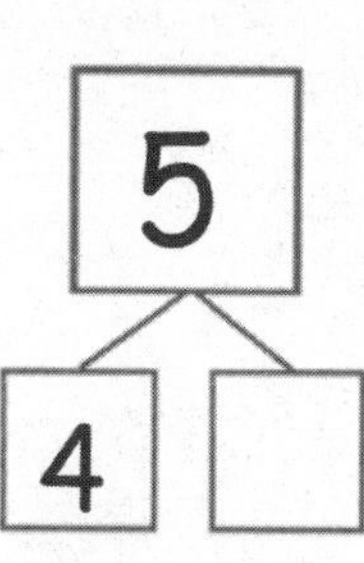

4.
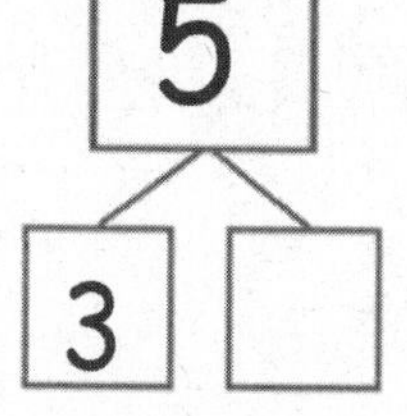

5.
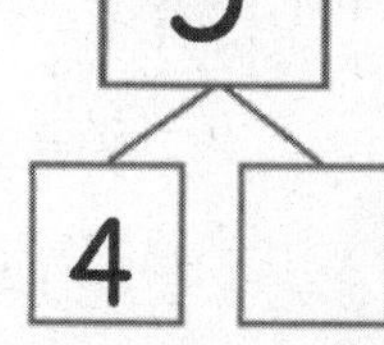

6.
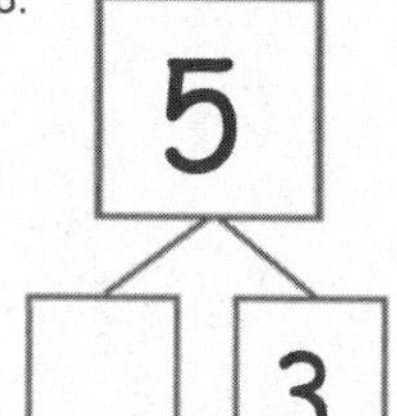

7.
5
2

8.
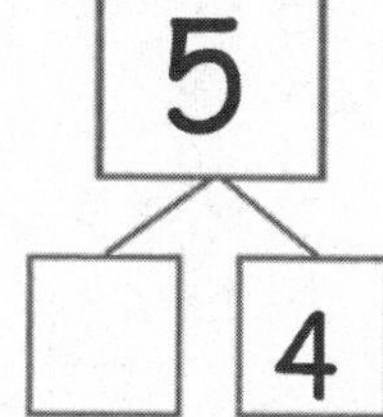

9.
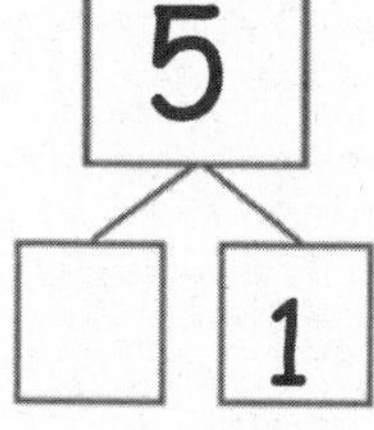

10.
5
2

11.
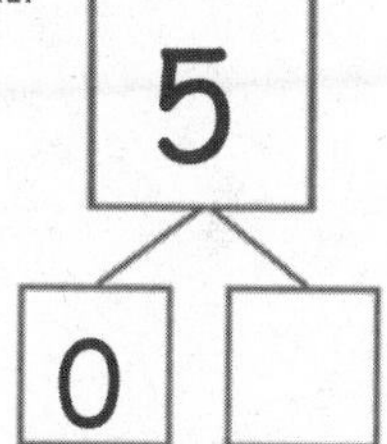

12.
5
1

13.
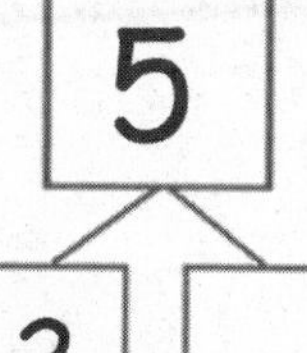

14.
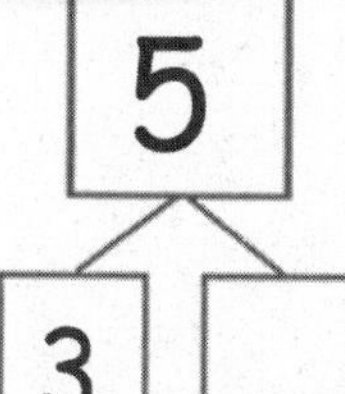

15.
5
4

16.
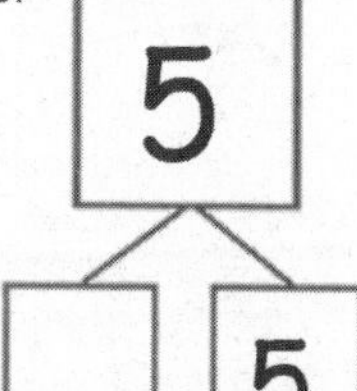

17.
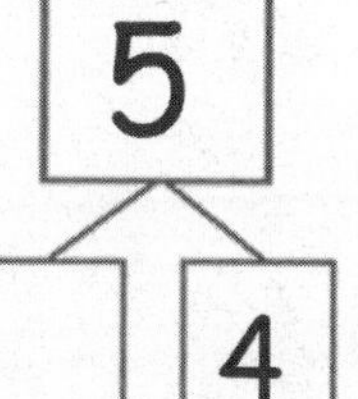

18.
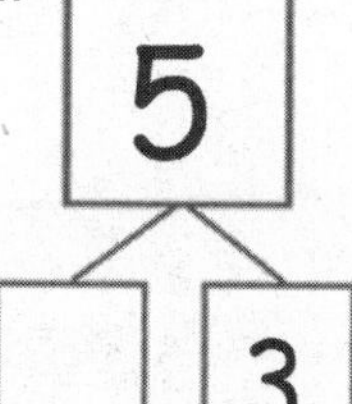

19.
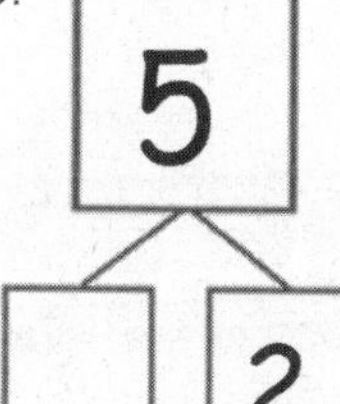

20.
5
1

21.

22.
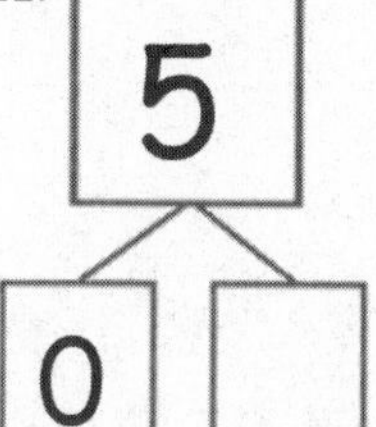

23.
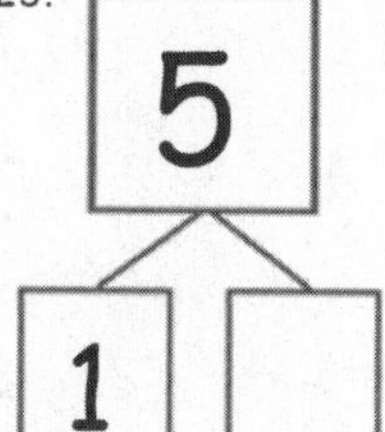

24.
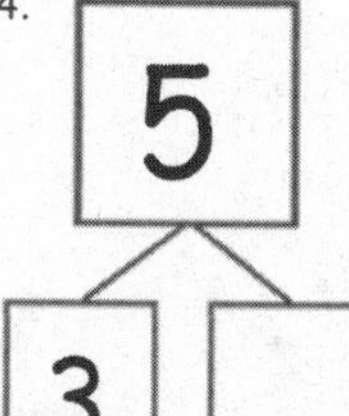

25.
5
2

____

carrera de vínculo numérico 5

A

Respuestas correctas:

Nombre ______________________ Fecha ____________

*Escribe el número que es 1 más.

1.	●●●		16.	●●●●● ●●●●	
2.	●●		17.	9	
3.	●●●		18.	7	
4.	●●●●		19.	●●●●● ●●	
5.	●●●●●		20.	8	
6.	●●●●● ●		21.	7	
7.	●●●●●		22.	●●●●● ●●●	
8.	5		23.	●●●●● ●●●●	
9.	●●●●● ●●		24.	10	
10.	6		25.	●●●●● ●●●●●	
11.	●●●●● ●		26.	●●●●● ●●●	
12.	7		27.	●● ●● ●● ●●	
13.	●●●●● ●●		28.	9	
14.	●●●●● ●●●		29.	●●● ●●● ●●●	
15.	8		30.	●●● ●●● ●●● ●●●	

B

Respuestas correctas:

Nombre ______________________ Fecha ____________

*Escribe el número que es 1 más.

1.	●●		16.	●●●●● ●●●	
2.	●		17.	8	
3.	●●		18.	9	
4.	●●●		19.	●●●●● ●●●●	
5.	●●●●		20.	●●●●● ●●●●●	
6.	●●●●●		21.	10	
7.	●●●●		22.	●●●●● ●●●	
8.	4		23.	●●●●● ●●●●	
9.	●●●●●		24.	10	
10.	5		25.	●●●●● ●●●●	
11.	●●●●●		26.	●● ●● ● ●● ●●	
12.	7		27.	●● ●● ●● ●●	
13.	●●●●● ●●		28.	8	
14.	●●●●● ●		29.	●● ●● ●● ●●●	
15.	6		30.	●●● ●●●● ●● ●●●●	

¡Agita esos discos!—6

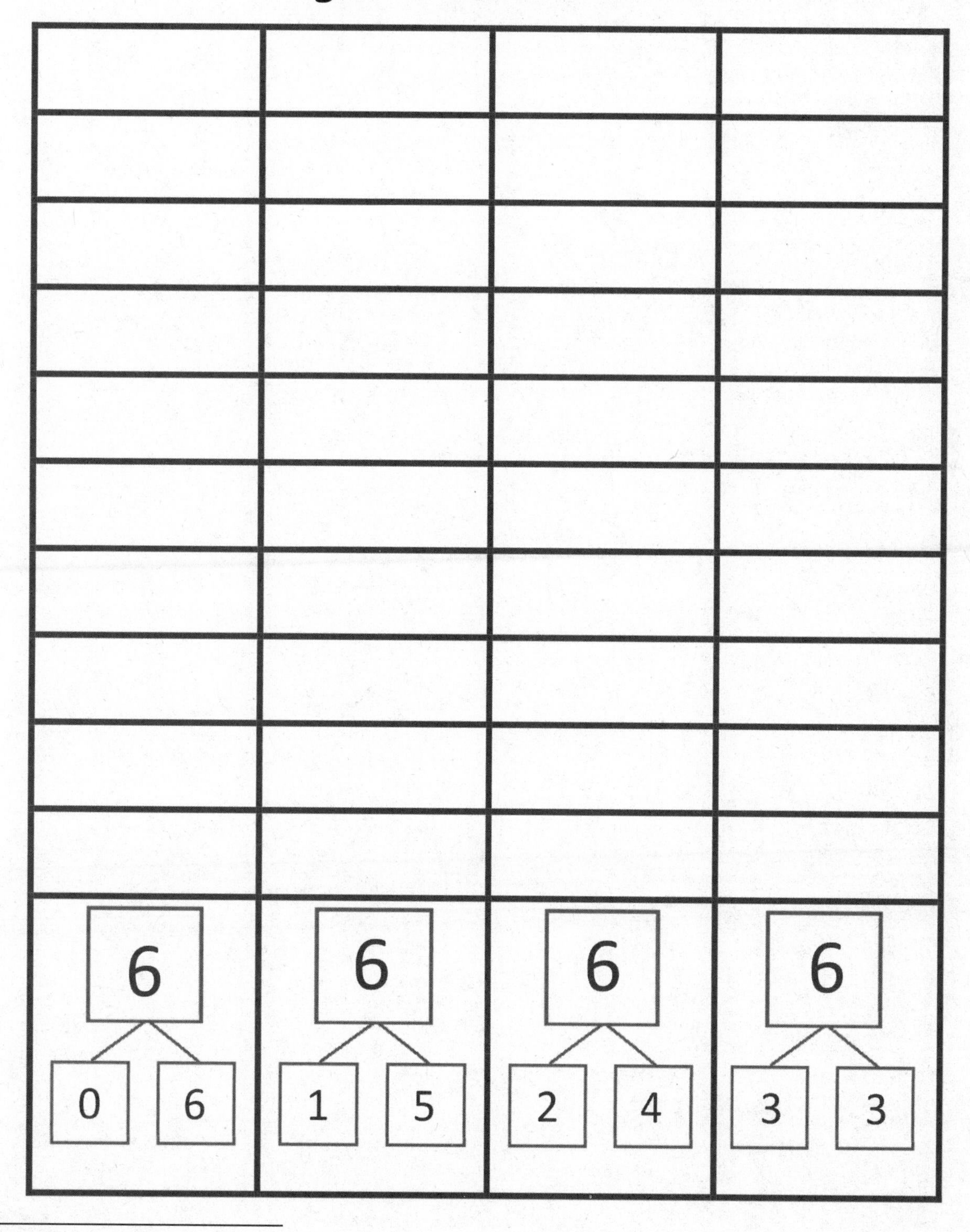

tablero de agita esos discos 6

Nombre ______________________________ Fecha __________

Haz tantos como puedas en 90 segundos. Escribe los vínculos numéricos que terminaste aquí:

1.
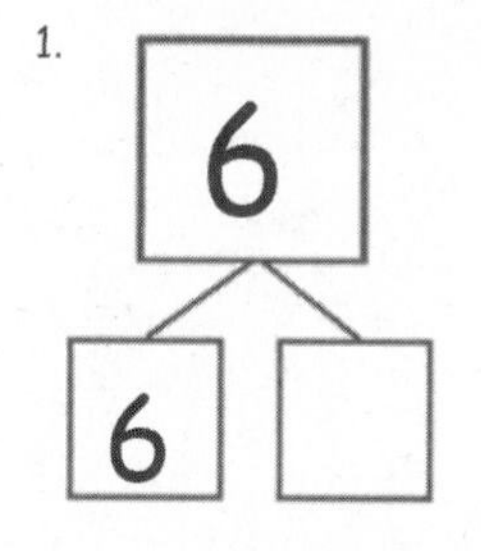

2. 6 — 5 ☐

3.
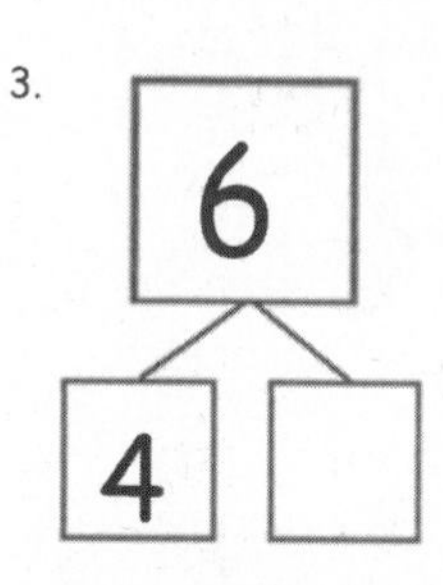

4.
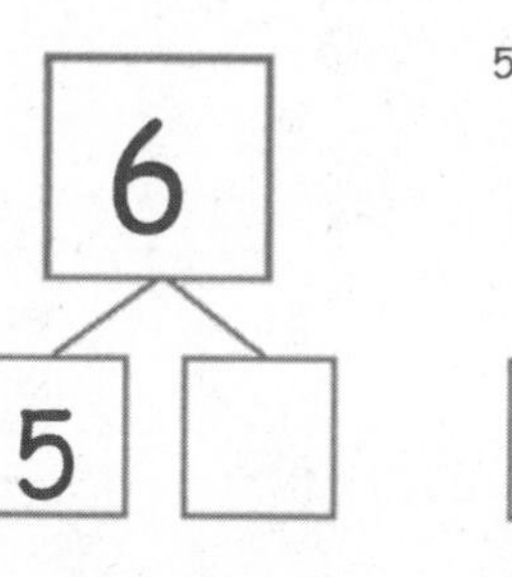

5.
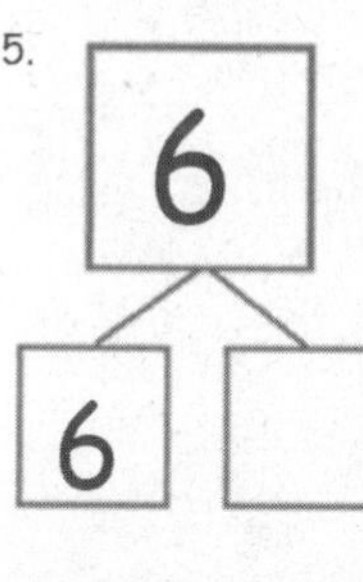

6.
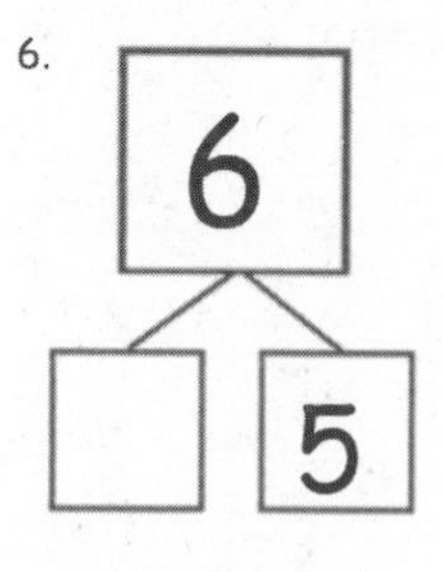

7.
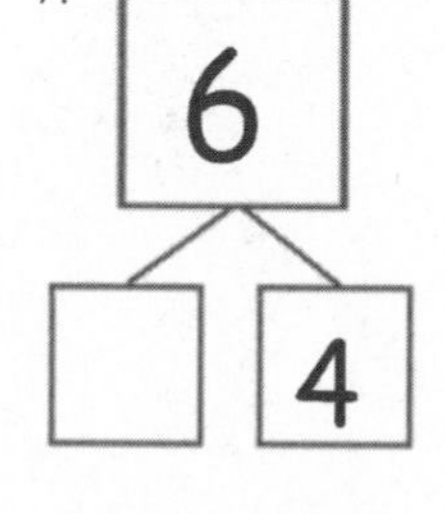

8.
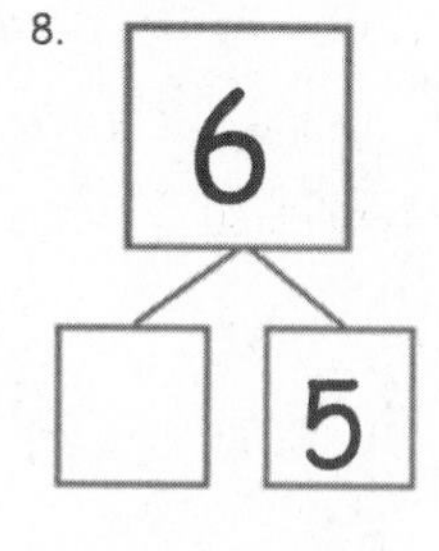

9.
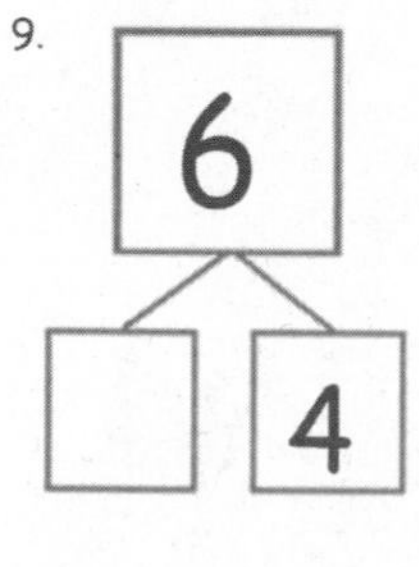

10.
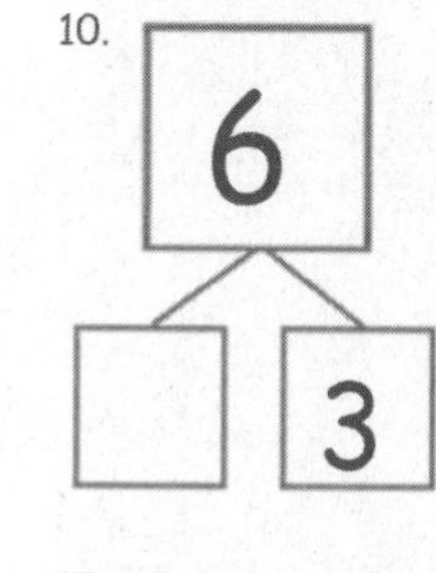

11.
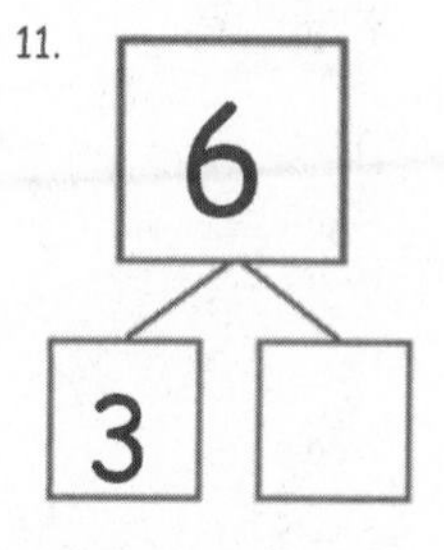

12.
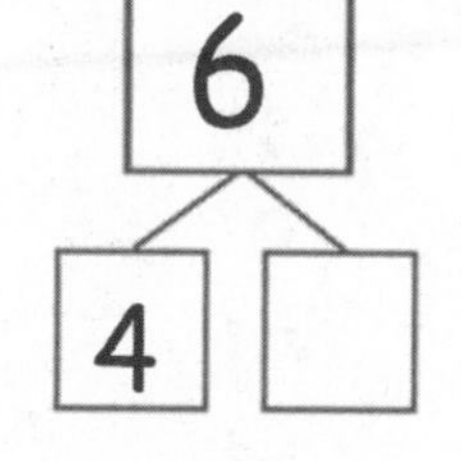

13.
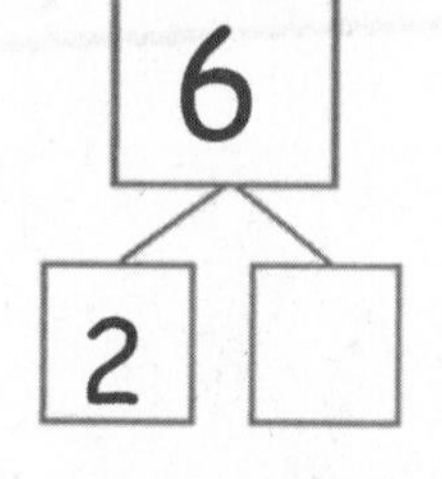

14.
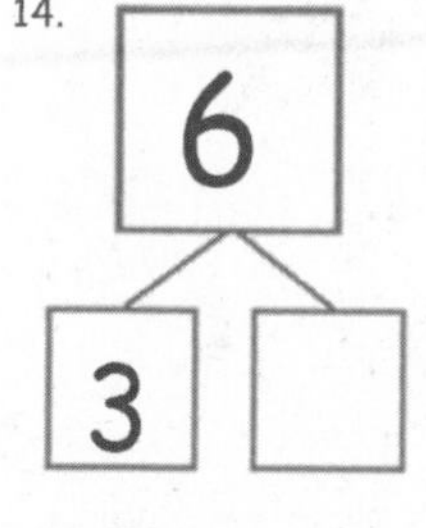

15.
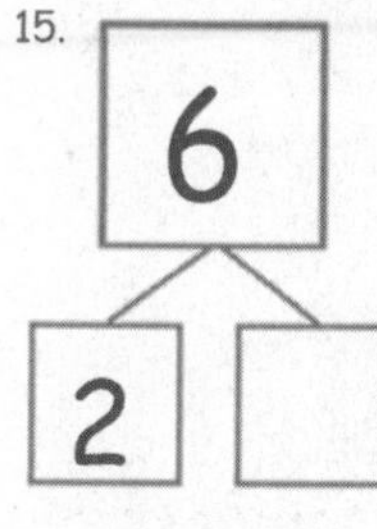

16.
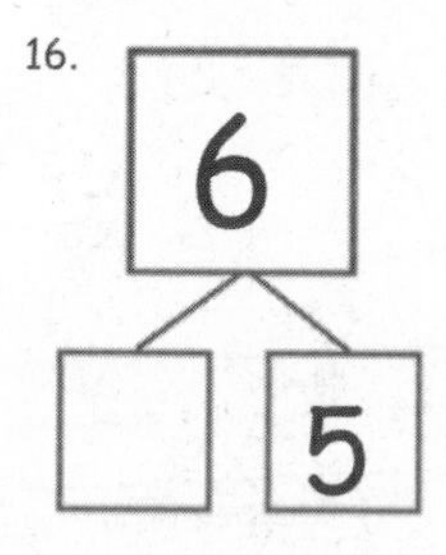

17.
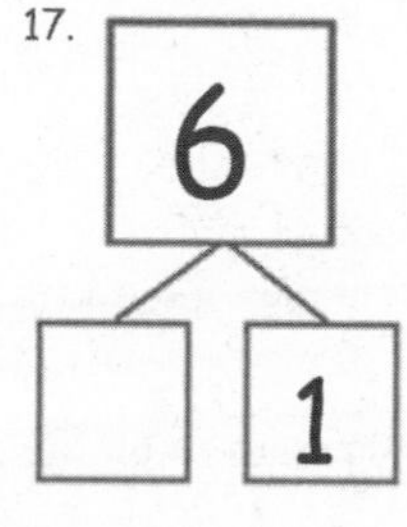

18.
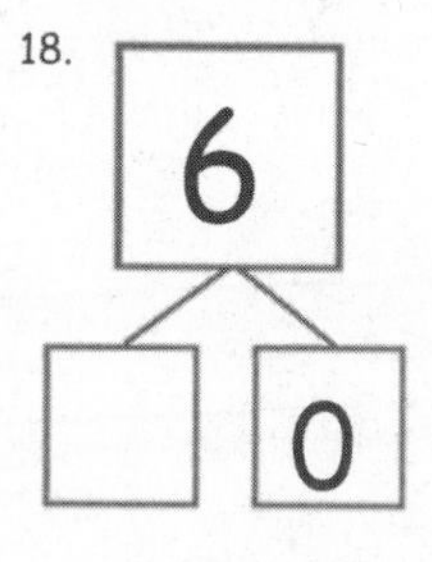

19.
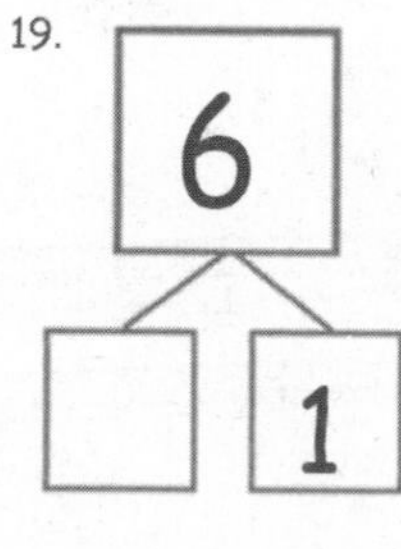

20.
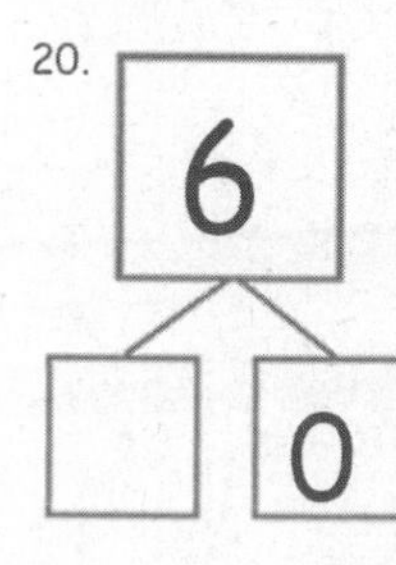

21. 6 — 1 ☐

22. 6 — 5 ☐

23.
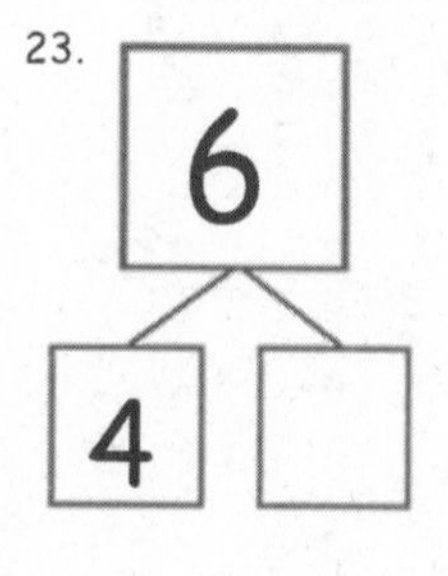

24. 6 — 2 ☐

25.
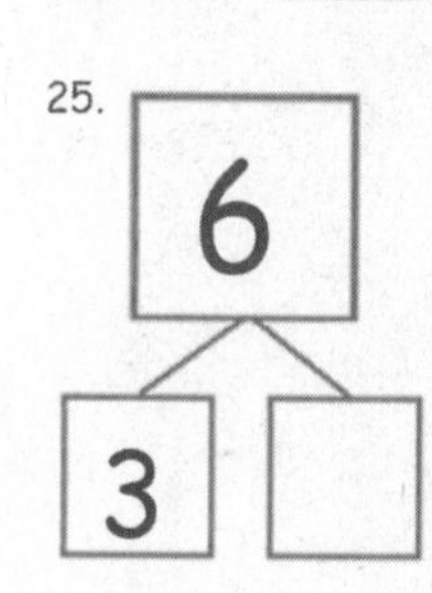

____

carrera de vínculos numéricos 6

Nombre ______________________________ Fecha __________

Haz tantos como puedas en 90 segundos. Escribe los vínculos numéricos que terminaste aquí:

Problema	Total	Parte	Parte
1.	7	6	
2.	7	7	
3.	7	6	
4.	7	5	
5.	7	6	
6.	7		7
7.	7		6
8.	7		5
9.	7		4
10.	7		3
11.	7	4	
12.	7	3	
13.	7	2	
14.	7	5	
15.	7	2	
16.	7		6
17.	7		1
18.	7		0
19.	7		2
20.	7		5
21.	7	1	
22.	7	5	
23.	7	3	
24.	7	0	
25.	7	6	

carrera de vínculo numérico 7

¡Agita esos discos!—8

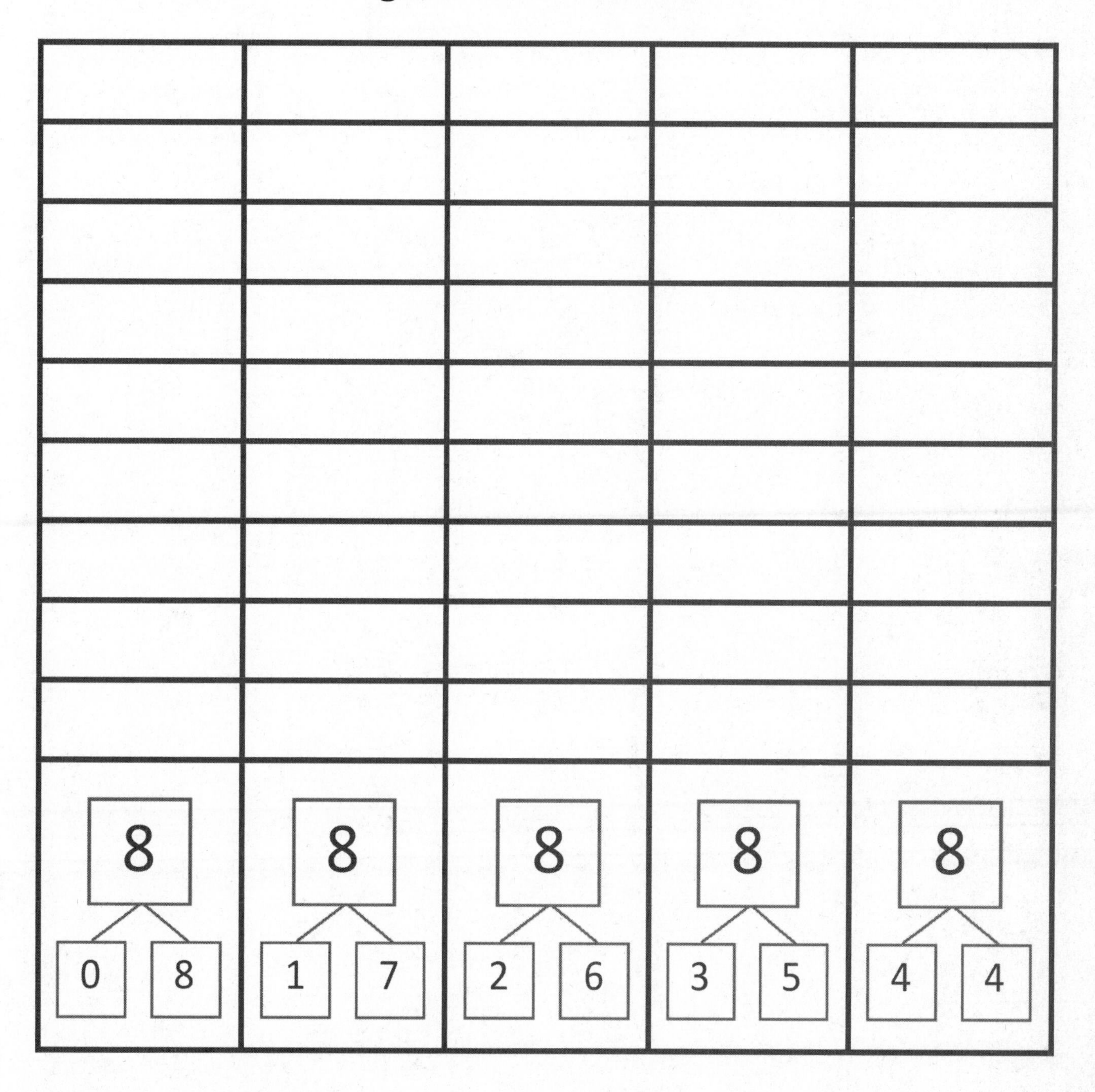

agita esos discos 8

Nombre ______________________ Fecha ________

Haz tantos como puedas en 90 segundos. Escribe los vínculos numéricos que terminaste aquí:

#	Total	Parte	Parte
1.	8	8	
2.	8	7	
3.	8	6	
4.	8	7	
5.	8	6	
6.	8		5
7.	8		6
8.	8		5
9.	8		4
10.	8		3
11.	8	4	
12.	8	5	
13.	8	3	
14.	8	4	
15.	8	3	
16.	8		6
17.	8		2
18.	8		6
19.	8		5
20.	8		3
21.	8	4	
22.	8	1	
23.	8	2	
24.	8	0	
25.	8	1	

carrera de vínculo numérico 8

Nombre ______________________ Fecha __________

Haz tantos como puedas en 90 segundos. Escribe cuántos vínculos numéricos que terminaste en la estrella.

Problema	Total	Parte	Parte
1.	9	8	
2.	9	7	
3.	9	8	
4.	9	7	
5.	9	9	
6.	9		6
7.	9		7
8.	9		6
9.	9		5
10.	9		4
11.	9	8	
12.	9	1	
13.	9	7	
14.	9	2	
15.	9	6	
16.	9		5
17.	9		6
18.	9		7
19.	9		2
20.	9		3
21.	9	5	
22.	9	1	
23.	9	2	
24.	9	0	
25.	9	2	

carrera de vínculos numéricos 9

Nombre ______________________________ Fecha __________

Haz lo más que puedas en 90 segundos. Escribe los vínculos numéricos que terminaste aquí:

1.
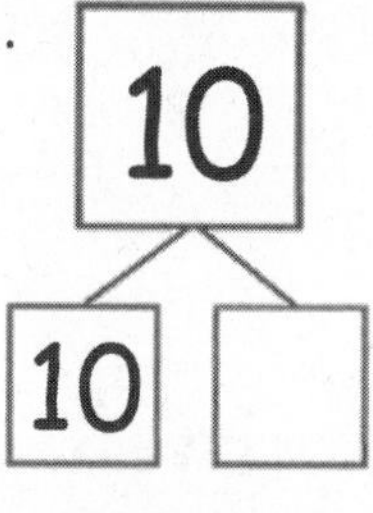

2.
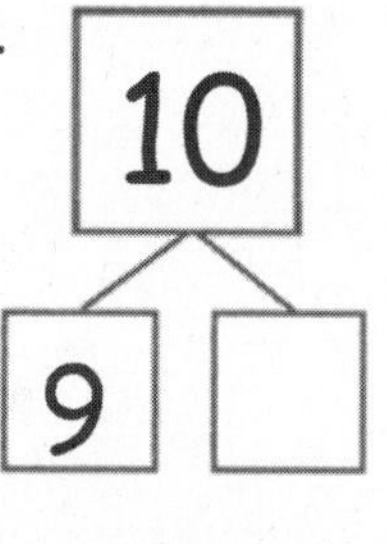

3.

4.
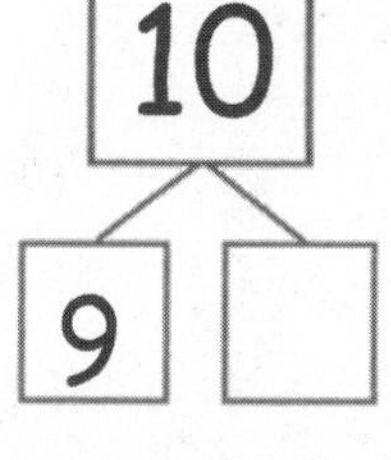

5.
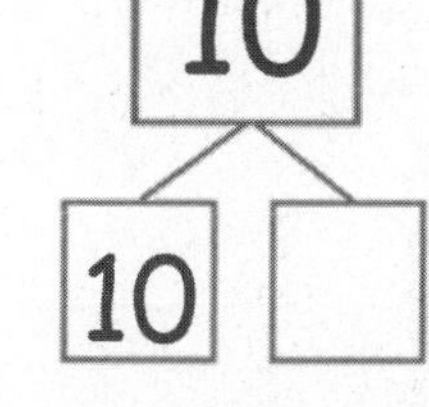

6.

7.
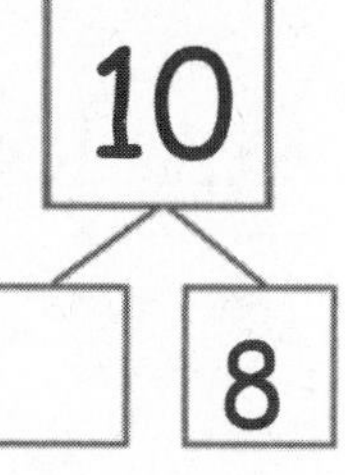

8.

9.
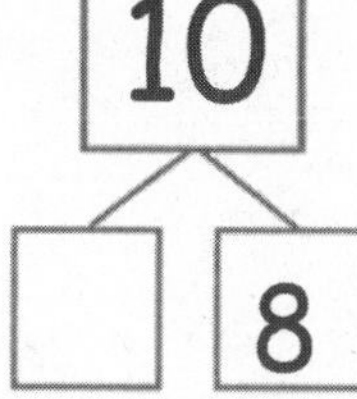

10.
10
7

11.

12.

13.

14.
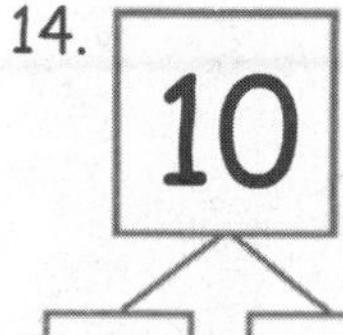

15.
10
4

16.

17.
10
4

18.

19.
10
4

20.
10
3

21.
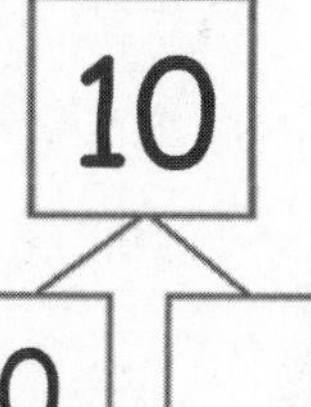

22.
10
1

23.
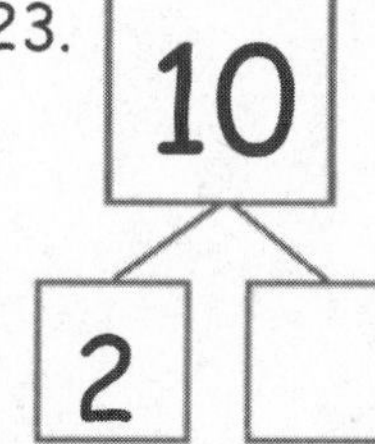

24.

25.
10
2

____

carrera de vínculos numéricos de 10

Número objetivo:

Ejercicio de tiro al blanco

Selecciona un *número objetivo* entre 6 y 10 y escríbelo en medio del círculo en la parte superior de la página. Tira un dado. Escribe el número que salío en uno de los círculos, en el extremo de las flechas. Luego da en el blanco escribiendo el número necesario en el otro círculo, para formar tu número objetivo.

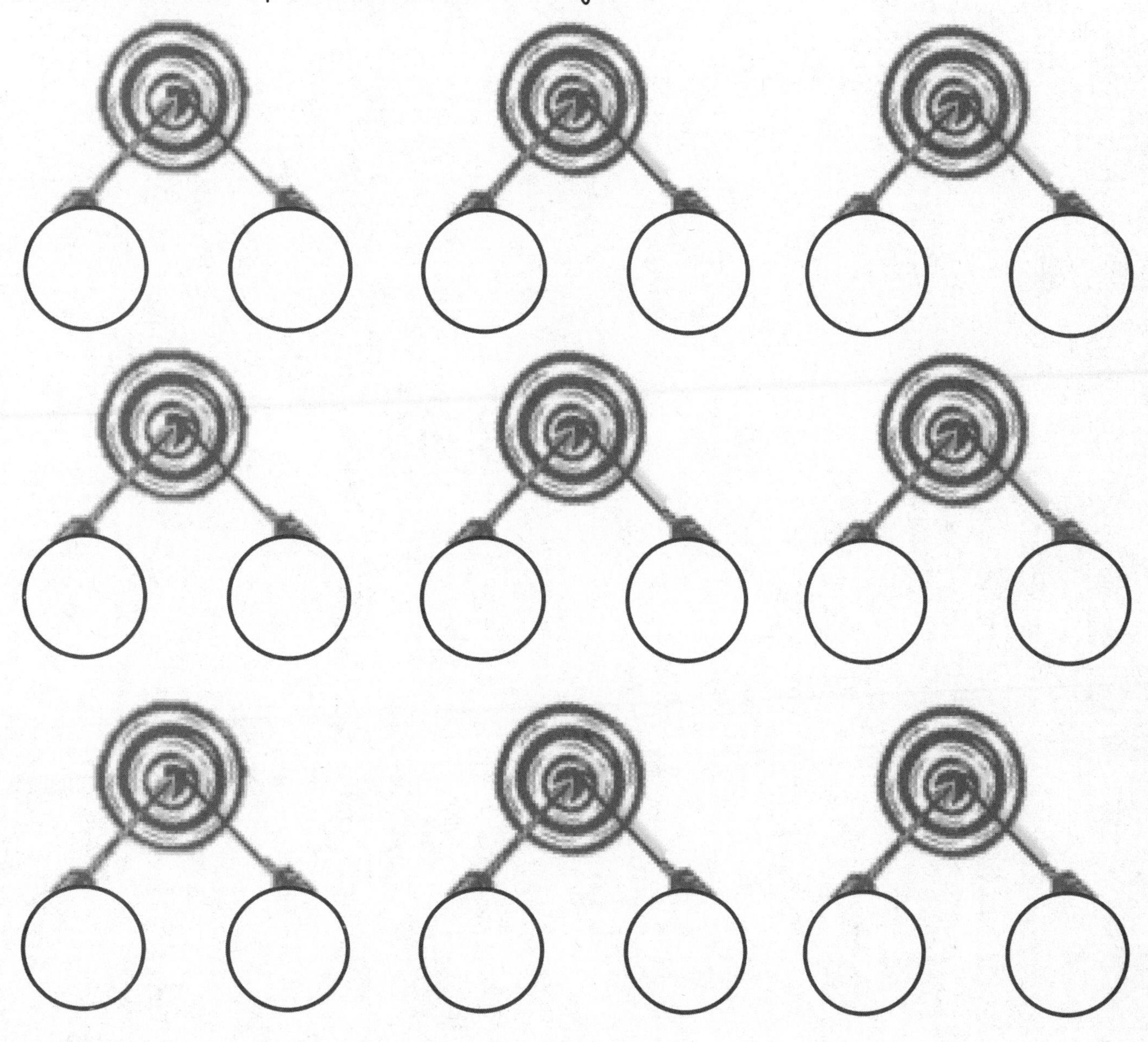

Ejercicio de tiro al blanco

A

Respuestas correctas:

Nombre ______________________ Fecha ______________

*Cuenta para sumar. Escribe el número.

1.	1 + 1 ● ●		16.	4 + 3 ●●●	
2.	2 + 1 ●● ●		17.	5 + 3 ●●●	
3.	3 + 1 ●●● ●		18.	7 + 3 ●●●	
4.	3 + 2 ●●● ●●		19.	7 + 2 ●●	
5.	1 + 2 ● ●●		20.	8 + 2 ●●	
6.	2 + 2 ●● ●●		21.	6 + 2 ●●	
7.	2 + 3 ●● ●●●		22.	6 + 1 ●	
8.	2 + 1 ●		23.	6 + 1	
9.	2 + 2 ●●		24.	6 + 2	
10.	3 + 2 ●●		25.	7 + 2	
11.	5 + 2 ●●		26.	8 + 2	
12.	8 + 2 ●●		27.	2 + 8	
13.	8 + 1 ●		28.	2 + 6	
14.	7 + 1 ●		29.	3 + 6	
15.	9 + 1 ●		30.	4 + 5	

B

Respuestas correctas:

Nombre ______________________ Fecha ______________

*Cuenta para sumar. Escribe el número.

1.	1 + 1		16.	4 + 2	
2.	2 + 2		17.	3 + 2	
3.	3 + 2		18.	5 + 2	
4.	2 + 2		19.	7 + 2	
5.	2 + 1		20.	7 + 3	
6.	3 + 1		21.	6 + 3	
7.	3 + 2		22.	6 + 2	
8.	3 + 2		23.	6 + 2	
9.	2 + 2		24.	5 + 2	
10.	4 + 2		25.	7 + 2	
11.	1 + 2		26.	6 + 2	
12.	2 + 1		27.	2 + 6	
13.	3 + 1		28.	2 + 7	
14.	5 + 1		29.	3 + 7	
15.	7 + 1		30.	4 + 7	

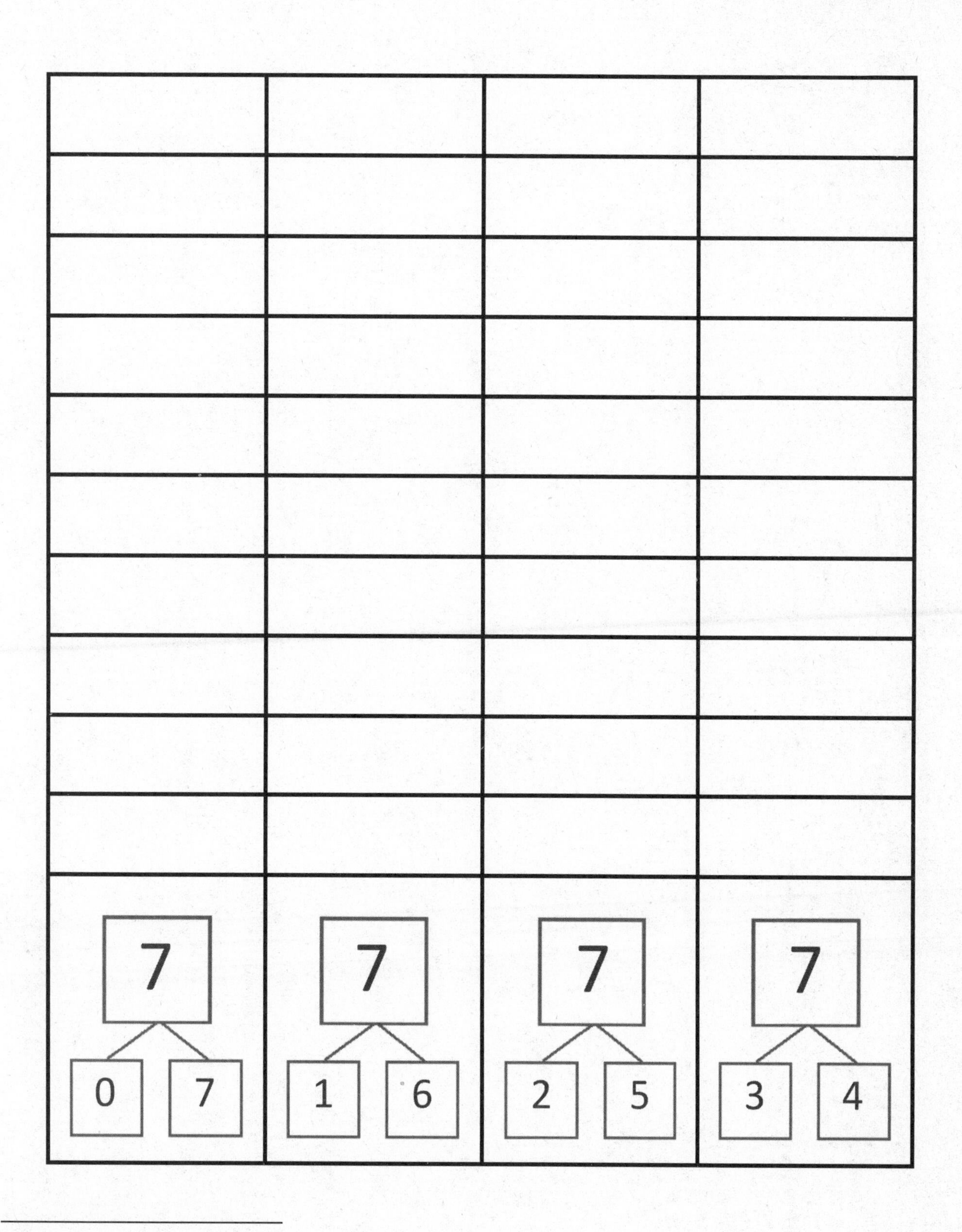

tabla agita esos discos 7

A

Respuestas correctas:

Nombre ______________________ Fecha ______________

*Cuenta para sumar.

1.	1 + 1		16.	4 + 3	
2.	2 + 1		17.	3 + 3	
3.	3 + 1		18.	4 + 3	
4.	3 + 2		19.	3 + 4	
5.	2 + 2		20.	2 + 4	
6.	3 + 2		21.	4 + 2	
7.	2 + 2		22.	5 + 2	
8.	3 + 0		23.	2 + 5	
9.	3 + 1		24.	2 + 6	
10.	3 + 2		25.	6 + 3	
11.	5 + 2		26.	3 + 6	
12.	5 + 3		27.	2 + 7	
13.	5 + 2		28.	3 + 7	
14.	5 + 3		29.	2 + 8	
15.	6 + 3		30.	3 + 6	

B

Respuestas correctas:

Nombre ______________________ Fecha ______________

*Cuenta para sumar.

1.	2 + 1		16.	4 + 3	
2.	1 + 1		17.	3 + 3	
3.	2 + 1		18.	2 + 3	
4.	2 + 2		19.	1 + 3	
5.	3 + 2		20.	0 + 3	
6.	2 + 2		21.	1 + 3	
7.	3 + 2		22.	2 + 5	
8.	3 + 1		23.	5 + 2	
9.	5 + 1		24.	2 + 6	
10.	6 + 1		25.	6 + 2	
11.	6 + 2		26.	3 + 6	
12.	5 + 2		27.	3 + 7	
13.	6 + 2		28.	2 + 7	
14.	6 + 3		29.	2 + 6	
15.	5 + 3		30.	3 + 6	

Nombre ________________________________ Fecha ________________

¡Carrera a la cima!

0	2	4	6	8	10

A

Respuestas correctas:

Nombre ______________________ Fecha ______________

*Escribe el número que es 1 menos.

1.	5		16.	10	
2.	4		17.	8	
3.	3		18.	11	
4.	5		19.	10	
5.	3		20.	9	
6.	1		21.	1	
7.	4		22.	11	
8.	5		23.	21	
9.	7		24.	4	
10.	6		25.	14	
11.	7		26.	24	
12.	9		27.	10	
13.	8		28.	20	
14.	9		29.	21	
15.	10		30.	31	

B

Respuestas correctas:

Nombre ______________________ Fecha ____________

*Escribe el número que es 1 menos.

1.	3		16.	10	
2.	2		17.	9	
3.	1		18.	11	
4.	6		19.	9	
5.	4		20.	13	
6.	2		21.	11	
7.	1		22.	1	
8.	3		23.	11	
9.	5		24.	21	
10.	7		25.	5	
11.	10		26.	15	
12.	9		27.	25	
13.	8		28.	20	
14.	6		29.	10	
15.	17		30.	21	

A

Respuestas correctas: ______

Suma.

1.	3 + 1 =	
2.	4 + 1 =	
3.	5 + 1 =	
4.	9 + 1 =	
5.	6 + 1 =	
6.	8 + 1 =	
7.	2 + 1 =	
8.	7 + 1 =	
9.	1 + 7 =	
10.	1 + 9 =	
11.	1 + 6 =	
12.	2 + 2 =	
13.	3 + 2 =	
14.	4 + 2 =	
15.	8 + 2 =	
16.	5 + 2 =	
17.	6 + 2 =	
18.	7 + 2 =	
19.	2 + 7 =	
20.	2 + 8 =	
21.	2 + 5 =	
22.	2 + 6 =	

23.	1 + 2 =	
24.	3 + 6 =	
25.	1 + 8 =	
26.	2 + 3 =	
27.	1 + 4 =	
28.	2 + 4 =	
29.	1 + 3 =	
30.	1 + 5 =	
31.	3 + 3 =	
32.	4 + 3 =	
33.	5 + 3 =	
34.	6 + 3 =	
35.	7 + 3 =	
36.	3 + 7 =	
37.	3 + 4 =	
38.	3 + 5 =	
39.	4 + 4 =	
40.	5 + 4 =	
41.	6 + 4 =	
42.	4 + 6 =	
43.	4 + 5 =	
44.	5 + 5 =	

B

Respuestas correctas: ______

Suma.

Mejora: ______

1.	2 + 1 =	
2.	3 + 1 =	
3.	4 + 1 =	
4.	8 + 1 =	
5.	5 + 1 =	
6.	7 + 1 =	
7.	9 + 1 =	
8.	6 + 1 =	
9.	1 + 6 =	
10.	1 + 9 =	
11.	1 + 7 =	
12.	2 + 2 =	
13.	3 + 2 =	
14.	4 + 2 =	
15.	7 + 2 =	
16.	5 + 2 =	
17.	8 + 2 =	
18.	6 + 2 =	
19.	2 + 6 =	
20.	2 + 8 =	
21.	2 + 5 =	
22.	2 + 7 =	

23.	1 + 8 =	
24.	3 + 7 =	
25.	1 + 5 =	
26.	2 + 4 =	
27.	1 + 4 =	
28.	2 + 3 =	
29.	1 + 3 =	
30.	1 + 2 =	
31.	3 + 3 =	
32.	4 + 3 =	
33.	5 + 3 =	
34.	7 + 3 =	
35.	6 + 3 =	
36.	3 + 6 =	
37.	3 + 5 =	
38.	3 + 4 =	
39.	4 + 4 =	
40.	5 + 4 =	
41.	6 + 4 =	
42.	4 + 6 =	
43.	4 + 5 =	
44.	5 + 5 =	

A

Respuestas correctas:

Nombre ______________________ Fecha ______________

*Escribe el número que falta de cada enunciado de resta. Presta atención al signo =.

1.	2 - 1 = □		16.	□ = 10 - 0	
2.	1 - 1 = □		17.	□ = 10 - 1	
3.	1 - 0 = □		18.	□ = 9 - 1	
4.	3 - 1 = □		19.	□ = 7 - 1	
5.	3 - 0 = □		20.	□ = 6 - 1	
6.	4 - 0 = □		21.	□ = 6 - 0	
7.	4 - 1 = □		22.	□ = 8 - 0	
8.	5 - 1 = □		23.	8 - □ = 8	
9.	6 - 1 = □		24.	□ - 0 = 8	
10.	6 - 0 = □		25.	7 - □ = 6	
11.	8 - 0 = □		26.	7 = 7 - □	
12.	10 - 0 = □		27.	9 = 9 - □	
13.	9 - 0 = □		28.	□ - 1 = 7	
14.	9 - 1 = □		29.	□ - 0 = 8	
15.	10 - 1 = □		30.	9 = □ - 1	

B

Respuestas correctas:

Nombre ______________________ Fecha ______________

*Escribe el número que falta de cada enunciado de resta. Presta atención al signo =.

1.	3 - 1 = □		16.	□ = 10 - 1	
2.	2 - 1 = □		17.	□ = 9 - 1	
3.	1 - 1 = □		18.	□ = 7 - 1	
4.	1 - 0 = □		19.	□ = 7 - 0	
5.	2 - 0 = □		20.	□ = 8 - 0	
6.	4 - 0 = □		21.	□ = 10 - 0	
7.	5 - 1 = □		22.	□ = 9 - 1	
8.	7 - 1 = □		23.	9 - □ = 8	
9.	8 - 1 = □		24.	□ - 1 = 8	
10.	9 - 0 = □		25.	7 - □ = 6	
11.	10 - 0 = □		26.	6 = 7 - □	
12.	7 - 0 = □		27.	9 = 9 - □	
13.	8 - 0 = □		28.	□ - 0 = 9	
14.	10 - 1 = □		29.	□ - 0 = 10	
15.	9 - 1 = □		30.	8 = □ - 1	

A

Respuestas correctas:

Nombre ______________________ Fecha ______________

Escribe el número que falta para cada enunciado de resta. Presta atención al signo =.

1.	2 - 2 = □		16.	0 = 10 - □	
2.	1 - 1 = □		17.	0 = 9 - □	
3.	1 - 0 = □		18.	0 = 8 - □	
4.	3 - 3 = □		19.	0 = 6 - □	
5.	3 - 2 = □		20.	1 = 6 - □	
6.	4 - 4 = □		21.	1 = 7 - □	
7.	4 - 3 = □		22.	1 = 10 - □	
8.	6 - 6 = □		23.	10 - □ = 1	
9.	7 - 7 = □		24.	□ - 9 = 1	
10.	8 - 8 = □		25.	7 - □ = 0	
11.	8 - 7 = □		26.	0 = 7 - □	
12.	9 - 9 = □		27.	0 = 9 - □	
13.	9 - 8 = □		28.	□ - 8 = 0	
14.	10 - 10 = □		29.	□ - 7 = 1	
15.	10 - 9 = □		30.	1 = □ - 5	

B

Respuestas correctas:

Nombre ______________________ Fecha ______________

Escribe el número que falta para cada enunciado de resta. Presta atención al signo =.

1.	3 - 3 = □		16.	0 = 6 - □	
2.	2 - 2 = □		17.	0 = 7 - □	
3.	1 - 1 = □		18.	0 = 8 - □	
4.	1 - 0 = □		19.	0 = 10 - □	
5.	2 - 1 = □		20.	1 = 10 - □	
6.	4 - 3 = □		21.	1 = 9 - □	
7.	5 - 4 = □		22.	1 = 7 - □	
8.	7 - 7 = □		23.	7 - □ = 1	
9.	8 - 8 = □		24.	□ - 6 = 1	
10.	9 - 9 = □		25.	6 - □ = 0	
11.	10 - 10 = □		26.	0 = 6 - □	
12.	10 - 9 = □		27.	0 = 8 - □	
13.	8 - 7 = □		28.	□ -8 = 0	
14.	6 - 5 = □		29.	□ -6 = 1	
15.	6 - 6 = □		30.	1 = □ - 6	

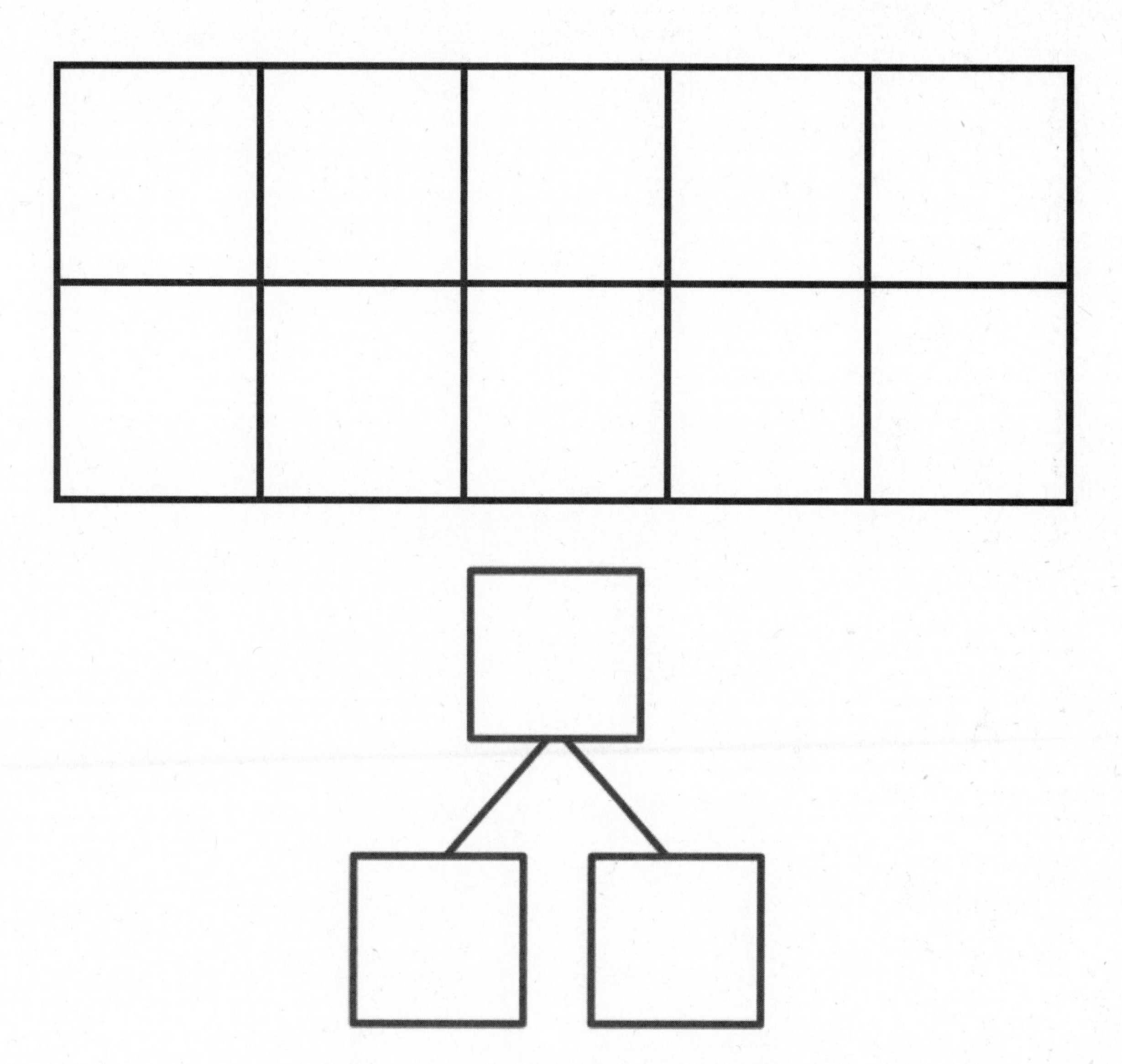

table de decenas

A

Respuestas correctas:

Nombre ______________________ Fecha ____________

*Escribe el número que falta para cada enunciado numérico. Presta atención a los signos de + y –.

1.	9 + 1 = □		16.	10 – 7 = □	
2.	1 + 9 = □		17.	10 = 7 + □	
3.	10 – 1 = □		18.	10 = 3 + □	
4.	10 – 9 = □		19.	10 = 6 + □	
5.	10 + 0 = □		20.	10 = 4 + □	
6.	0 + 10 = □		21.	10 = 5 + □	
7.	10 – 0 = □		22.	10 – □ = 5	
8.	10 – 10 = □		23.	5 = 10 – □	
9.	8 + 2 = □		24.	6 = 10 – □	
10.	2 + 8 = □		25.	7 = 10 – □	
11.	10 – 2 = □		26.	7 = □ – 3	
12.	10 – 8 = □		27.	4 = 10 – □	
13.	7 + 3 = □		28.	5 = □ – 5	
14.	3 + 7 = □		29.	6 = 10 – □	
15.	10 – 3 = □		30.	7 = □ – 3	

B

Respuestas correctas:

Nombre ______________________ Fecha ____________

*Escribe el número que falta para cada enunciado numérico. Presta atención a los signos de + y –.

1.	8 + 2 = □		16.	10 – 6 = □	
2.	2 + 8 = □		17.	10 = 8 + □	
3.	10 – 2 = □		18.	10 = 7 + □	
4.	10 – 8 = □		19.	10 = 3 + □	
5.	9 + 1 = □		20.	10 = 4 + □	
6.	1 + 9 = □		21.	10 = 5 + □	
7.	10 – 1 = □		22.	10 – □ = 5	
8.	10 – 9 = □		23.	6 = 10 – □	
9.	10 + 0 = □		24.	7 = 10 – □	
10.	0 + 10 = □		25.	8 = 10 – □	
11.	10 – 0 = □		26.	7 = □ – 3	
12.	10 – 10 = □		27.	2 = 10 – □	
13.	6 + 4 = □		28.	4 = □ – 6	
14.	4 + 6 = □		29.	3 = 10 – □	
15.	10 – 4 = □		30.	7 = □ – 3	

A

Respuestas correctas:

Nombre ______________________ Fecha ______________

*Escribe el número que falta en cada enunciado.

1.	8 y 2 suman ☐		16.	11 es 10 y ☐	
2.	9 y 1 suman ☐		17.	11 es 1 y ☐	
3.	7 y 3 suman ☐		18.	12 es 2 y ☐	
4.	6 y ☐ suman 10		19.	11 es ☐ y 1	
5.	4 y ☐ suman 10		20.	14 es 10 y ☐	
6.	5 e ☐ suman 10		21.	15 es 5 y ☐	
7.	☐ y 5 suman 10		22.	18 es 8 y ☐	
8.	13 es 10 y ☐		23.	20 es 10 y ☐	
9.	14 es 10 y ☐		24.	2 más 10 es ☐	
10.	16 es 10 y ☐		25.	10 más 2 es ☐	
11.	17 es 10 y ☐		26.	10 es ☐ menos 12	
12.	19 es 10 y ☐		27.	10 es ☐ menos 12	
13.	18 es 10 y ☐		28.	8 menos 18 es ☐	
14.	12 es 10 y ☐		29.	6 menos 16 es ☐	
15.	13 es 10 y ☐		30.	10 menos 20 es ☐	

B

Respuestas correctas:

Nombre ______________________ Fecha ____________

*Escribe el número que falta en cada enunciado.

1.	9 y 1 suman ☐		16.	13 es 10 y ☐	
2.	8 y 2 suman ☐		17.	13 es 3 y ☐	
3.	6 y 4 suman ☐		18.	11 es 1 y ☐	
4.	7 y ☐ suman 10		19.	11 es ☐ y 1	
5.	3 y ☐ suman 10		20.	15 es ☐ y 10	
6.	4 y ☐ suman 10		21.	14 es 4 y ☐	
7.	☐ y 5 suman 10		22.	19 es 9 y ☐	
8.	14 es 10 y ☐		23.	20 es 10 y ☐	
9.	13 es 10 y ☐		24.	1 más 10 es ☐	
10.	17 es 10 y ☐		25.	10 más 1 es ☐	
11.	16 es 10 y ☐		26.	10 es ☐ menos 11	
12.	15 es 10 y ☐		27.	10 es ☐ menos 14	
13.	19 es 10 y ☐		28.	7 menos 18 es ☐	
14.	11 es 10 y ☐		29.	7 menos 16 es ☐	
15.	12 es 10 y ☐		30.	10 menos 20 es ☐	

1.er grado
Módulo 2

A

Respuestas correctas:

Nombre ______________________________ Fecha ______________

*Haz una decena para sumar.

1.	$9 + 1 + 3 = \square$		16.	$6 + 4 + 5 = \square$	
2.	$9 + 1 + 5 = \square$		17.	$6 + 4 + 6 = \square$	
3.	$1 + 9 + 5 = \square$		18.	$4 + 6 + 6 = \square$	
4.	$1 + 9 + 1 = \square$		19.	$4 + 6 + 5 = \square$	
5.	$5 + 5 + 4 = \square$		20.	$4 + 5 + 6 = \square$	
6.	$5 + 5 + 6 = \square$		21.	$5 + 3 + 5 = \square$	
7.	$5 + 5 + 5 = \square$		22.	$6 + 5 + 5 = \square$	
8.	$8 + 2 + 1 = \square$		23.	$1 + 4 + 9 = \square$	
9.	$8 + 2 + 3 = \square$		24	$9 + 1 + \square = 14$	
10.	$8 + 2 + 7 = \square$		25.	$8 + 2 + \square = 11$	
11.	$2 + 8 + 7 = \square$		26.	$\square + 3 + 4 = 13$	
12.	$7 + 3 + 3 = \square$		27.	$2 + \square + 6 = 16$	
13.	$7 + 3 + 6 = \square$		28.	$1 + 1 + \square = 11$	
14.	$7 + 3 + 7 = \square$		29.	$19 = 5 + \square + 9$	
15.	$3 + 7 + 7 = \square$		30.	$18 = 2 + \square + 6$	

B

Respuestas correctas:

Nombre ______________________ Fecha ______________

*Haz una decena para sumar.

1.	5 + 5 + 4 = □		16.	6 + 4 + 2 = □	
2.	5 + 5 + 6 = □		17.	6 + 4 + 3 = □	
3.	5 + 5 + 5 = □		18.	4 + 6 + 3 = □	
4.	9 + 1 + 1 = □		19.	4 + 6 + 6 = □	
5.	9 + 1 + 2 = □		20.	4 + 7 + 6 = □	
6.	9 + 1 + 5 = □		21.	5 + 4 + 5 = □	
7.	1 + 9 + 5 = □		22.	8 + 5 + 5 = □	
8.	1 + 9 + 6 = □		23.	1 + 7 + 9 = □	
9.	8 + 2 + 4 = □		24.	9 + 1 + □ = 11	
10.	8 + 2 + 7 = □		25.	8 + 2 + □ = 12	
11.	2 + 8 + 7 = □		26.	□ + 3 + 4 = 14	
12.	7 + 3 + 7 = □		27.	3 + □ + 7 = 20	
13.	7 + 3 + 8 = □		28.	7 + 8 + □ = 17	
14.	7 + 3 + 9 = □		29.	16 = 3 + □ + 6	
15.	3 + 7 + 9 = □		30.	19 = 2 + □ + 7	

A

Respuestas correctas:

Nombre ______________________ Fecha ______________

*Escribe el número faltante.

1.	9 + 1 = □		16.	9 + 5 = □	
2.	10 + 1 = □		17.	9 + 6 = □	
3.	9 + 2 = □		18.	6 + 9 = □	
4.	9 + 1 = □		19.	9 + 4 = □	
5.	10 + 2 = □		20.	4 + 9 = □	
6.	9 + 3 = □		21.	9 + 8 = □	
7.	9 + 1 = □		22.	9 + 9 = □	
8.	10 + 4 = □		23.	9 + □ = 18	
9.	9 + 5 = □		24.	□ + 6 = 15	
10.	9 + 1 = □		25.	□ + 6 = 16	
11.	10 + 6 = □		26.	13 = 9 + □	
12.	9 + 7 = □		27.	17 = 8 + □	
13.	9 + 1 = □		28.	10 + 2 = 9 + □	
14.	10 + 8 = □		29.	9 + 5 = 10 + □	
15.	9 + 9 = □		30.	□ + 7 = 8 + 9	

B

Respuestas correctas:

Nombre ______________________ Fecha ______________

*Escribe el número faltante.

1.	9 + 1 = □		16.	5 + 9 = □	
2.	10 + 2 = □		17.	6 + 9 = □	
3.	9 + 3 = □		18.	9 + 6 = □	
4.	9 + 1 = □		19.	9 + 7 = □	
5.	10 + 1 = □		20.	7 + 9 = □	
6.	9 + 2 = □		21.	9 + 8 = □	
7.	9 + 1 = □		22.	9 + 9 = □	
8.	10 + 3 = □		23	9 + □ = 17	
9.	9 + 4 = □		24.	□ + 5 = 14	
10.	9 + 1 = □		25.	□ + 4 = 14	
11.	10 + 5 = □		26.	15 = 9 + □	
12.	9 + 6 = □		27.	16 = 7 + □	
13.	9 + 1 = □		28.	10 + 4 = 9 + □	
14.	10 + 4 = □		29.	9 + 6 = 10 + □	
15.	9 + 5 = □		30.	□ + 6 = 7 + 9	

A

Respuestas correctas:

Nombre ______________________ Fecha ______________

*Escribe el número faltante.

1.	$9 + 2 = \square$		16.	$4 + 8 = \square$	
2.	$9 + 3 = \square$		17	$8 + 4 = \square$	
3.	$9 + 5 = \square$		18.	$7 + 4 = \square$	
4.	$9 + 4 = \square$		19.	$7 + 5 = \square$	
5.	$8 + 2 = \square$		20.	$7 + 6 = \square$	
6.	$8 + 3 = \square$		21.	$6 + 7 = \square$	
7.	$8 + 5 = \square$		22.	$9 + 9 = \square$	
8.	$8 + 4 = \square$		23.	$9 + \square = 18$	
9.	$9 + 4 = \square$		24.	$\square + 4 = 13$	
10.	$8 + 5 = \square$		25.	$\square + 4 = 12$	
11	$9 + 5 = \square$		26.	$12 = 3 + \square$	
12.	$8 + 6 = \square$		27.	$16 = 8 + \square$	
13.	$9 + 6 = \square$		28	$9 + 4 = 8 + \square$	
14.	$6 + 9 = \square$		29.	$9 + 3 = 5 + \square$	
15.	$9 + 6 = \square$		30.	$\square + 7 = 8 + 6$	

B

Respuestas correctas:

Nombre ______________________ Fecha ______________

*Escribe el número faltante.

1.	$9 + 1 = \square$		16.	$3 + 8 = \square$	
2.	$9 + 2 = \square$		17	$8 + 3 = \square$	
3.	$9 + 4 = \square$		18.	$7 + 3 = \square$	
4.	$9 + 3 = \square$		19.	$7 + 4 = \square$	
5.	$8 + 2 = \square$		20.	$7 + 5 = \square$	
6.	$8 + 3 = \square$		21.	$5 + 7 = \square$	
7.	$8 + 5 = \square$		22.	$8 + 8 = \square$	
8.	$8 + 4 = \square$		23.	$8 + \square = 16$	
9.	$9 + 4 = \square$		24.	$\square + 3 = 12$	
10.	$8 + 5 = \square$		25.	$\square + 4 = 12$	
11	$9 + 5 = \square$		26.	$12 = 3 + \square$	
12.	$8 + 7 = \square$		27.	$14 = 7 + \square$	
13.	$9 + 7 = \square$		28	$9 + 3 = 8 + \square$	
14.	$7 + 9 = \square$		29.	$9 + 3 = 5 + \square$	
15.	$9 + 7 = \square$		30.	$\square + 7 = 8 + 5$	

OOOOO OOOOO

inserción de fila de grupos de 5

A

Respuestas correctas:

Nombre ____________________ Fecha ____________

*Escribe el número faltante.

1.	10 - 9 = □		16.	10 - □ = 5	
2.	10 - 8 = □		17.	9 - □ = 5	
3.	10 - 6 = □		18.	8 - □ = 5	
4.	10 - 7 = □		19.	10 - □ = 3	
5.	10 - 6 = □		20.	9 - □ = 3	
6.	10 - 5 = □		21.	8 - □ = 3	
7.	10 - 6 = □		22.	□ - 6 = 4	
8.	10 - 4 = □		23.	□ - 6 = 3	
9.	10 - 3 = □		24.	□ - 6 = 2	
10.	10 - 7 = □		25.	10 - 4 = 9 - □	
11.	10 - 8 = □		26.	8 - 2 = 10 - □	
12.	10 - 2 = □		27.	8 - □ = 10 - 3	
13.	10 - 1 = □		28.	9 - □ = 10 - 3	
14.	10 - 9 = □		29.	10 - 4 = 9 - □	
15.	10 - 10 = □		30.	□ - 2 = 10 - 4	

B

Respuestas correctas:

Nombre ______________________ Fecha ______________

*Escribe el número faltante.

1.	10 - 8 = □		16.	10 - □ = 0	
2.	10 - 9 = □		17.	9 - □ = 0	
3.	10 - 8 = □		18.	8 - □ = 0	
4.	10 - 9 = □		19.	10 - □ = 1	
5.	10 - 7 = □		20.	9 - □ = 1	
6.	10 - 9 = □		21.	8 - □ = 1	
7.	10 - 8 = □		22.	□ - 5 = 5	
8.	10 - 7 = □		23.	□ - 5 = 4	
9.	10 - 3 = □		24.	□ - 5 = 3	
10.	10 - 7 = □		25.	10 - 8 = 9 - □	
11.	10 - 6 = □		26.	8 - 6 = 10 - □	
12.	10 - 4 = □		27.	8 - □ = 10 - 2	
13.	10 - 3 = □		28.	9 - □ = 10 - 2	
14.	10 - 7 = □		29.	10 - 3 = 9 - □	
15.	10 - 5 = □		30.	□ - 1 = 10 - 3	

A

Respuestas correctas:

Nombre ______________________ Fecha ____________

*Escribe el número faltante. Presta atención al signo de sumar o restar.

1.	10 - 9 = □		16.	10 - 9 = □	
2.	1 + 2 = □		17.	11 - 9 = □	
3.	10 - 9 = □		18.	12 - 9 = □	
4.	1 + 3 = □		19.	15 - 9 = □	
5.	10 - 9 = □		20.	14 - 9 = □	
6.	1 + 1 = □		21.	13 - 9 = □	
7.	10 - 9 = □		22.	17 - 9 = □	
8.	1 + 2 = □		23.	18 - 9 = □	
9.	12 - 9 = □		24.	9 + □ = 13	
10.	10 - 9 = □		25.	9 + □ = 14	
11.	1 + 3 = □		26.	9 + □ = 16	
12.	13 - 9 = □		27.	9 + □ = 15	
13.	10 - 9 = □		28.	9 + □ = 17	
14.	1 + 5 = □		29.	9 + □ = 18	
15.	15 - 9 = □		30.	9 + □ = 19	

B

Respuestas correctas:

Nombre ______________________ Fecha ______________

*Escribe el número faltante. Presta atención al signo de suma o resta.

1.	10 - 9 = □		16.	10 - 9 = □	
2.	1 + 1 = □		17.	11 - 9 = □	
3.	10 - 9 = □		18.	13 - 9 = □	
4.	1 + 2 = □		19.	14 - 9 = □	
5.	10 - 9 = □		20.	13 - 9 = □	
6.	1 + 3 = □		21.	12 - 9 = □	
7.	10 - 9 = □		22.	15 - 9 = □	
8.	1 + 4 = □		23.	16 - 9 = □	
9.	14 - 9 = □		24.	9 + □= 12	
10.	10 - 9 = □		25.	9 + □ = 13	
11.	1 + 3 = □		26.	9 + □ = 15	
12.	13 - 9 = □		27.	9 + □ = 14	
13.	10 - 9 = □		28.	9 + □ = 15	
14.	1 + 2 = □		29.	9 + □ = 17	
15.	12 - 9 = □		30.	9 + □ = 16	

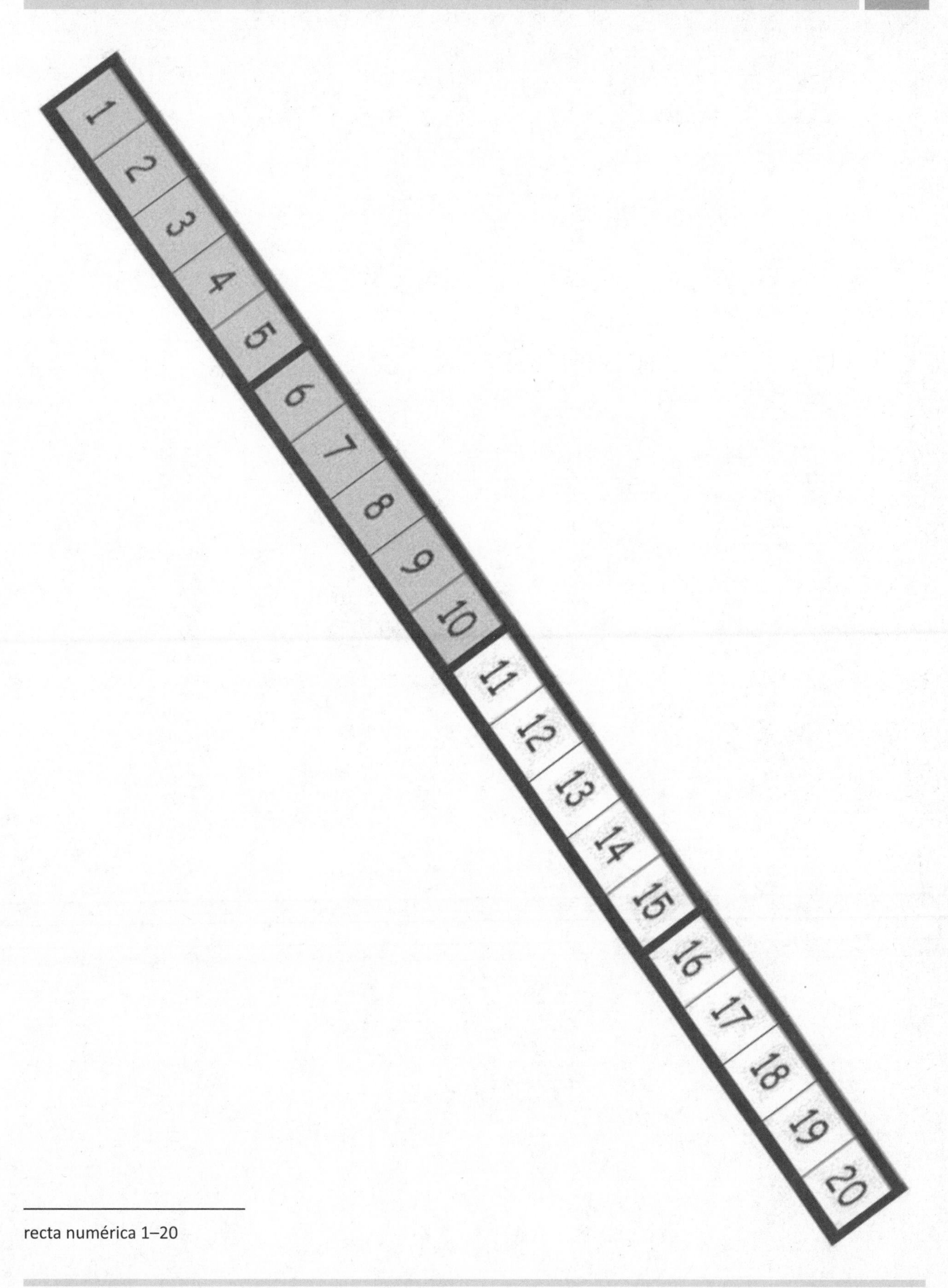

recta numérica 1–20

A

Respuestas correctas:

Nombre ____________________ Fecha ____________

*Escribe el número faltante. Presta atención al signo de suma o resta.

1.	10 - 8 = □		16.	10 - 8 = □	
2.	2 + 2 = □		17.	11 - 8 = □	
3.	10 - 8 = □		18.	12 - 8 = □	
4.	2 + 3 = □		19.	15 - 8 = □	
5.	10 - 8 = □		20.	14 - 8 = □	
6.	2 + 4 = □		21.	13 - 8 = □	
7.	10 - 8 = □		22.	17 - 8 = □	
8.	2 + 1 = □		23.	18 - 8 = □	
9.	11 - 8 = □		24.	8 + □ = 11	
10.	10 - 8 = □		25.	8 + □ = 12	
11.	2 + 2 = □		26.	8 + □ = 15	
12.	12 - 8 = □		27.	8 + □ = 14	
13.	10 - 8 = □		28.	8 + □ = 16	
14.	2 + 5 = □		29.	8 + □ = 17	
15.	15 - 8 = □		30.	8 + □ = 18	

B

Respuestas correctas:

Nombre ______________________ Fecha ____________

*Escribe el número faltante. Presta atención al signo de suma o resta.

1.	10 – 8 = □		16.	10 – 8 = □	
2.	2 + 1 = □		17.	11 – 8 = □	
3.	10 – 8 = □		18.	13 – 8 = □	
4.	2 + 2 = □		19.	14 – 8 = □	
5.	10 – 8 = □		20.	13 – 8 = □	
6.	2 + 3 = □		21.	12 – 8 = □	
7.	10 – 8 = □		22.	15 – 8 = □	
8.	2 + 2 = □		23.	16 – 8 = □	
9.	12 – 8 = □		24.	8 + □ = 10	
10.	10 – 8 = □		25.	8 + □ = 11	
11.	2 + 3 = □		26.	8 + □ = 13	
12.	13 – 8 = □		27.	8 + □ = 12	
13.	10 – 8 = □		28.	8 + □ = 13	
14.	2 + 2 = □		29.	8 + □ = 15	
15.	12 – 8 = □		30.	8 + □ = 16	

A

Respuestas correctas:

Nombre ______________________ Fecha ______________

*Escribe el número faltante.

1.	$10 - 9 = \square$		16.	$12 - 7 = \square$	
2.	$11 - 9 = \square$		17.	$13 - 7 = \square$	
3.	$13 - 9 = \square$		18.	$14 - 7 = \square$	
4.	$10 - 8 = \square$		19.	$15 - 9 = \square$	
5.	$11 - 8 = \square$		20.	$15 - 8 = \square$	
6.	$13 - 8 = \square$		21.	$15 - 7 = \square$	
7.	$10 - 7 = \square$		22.	$17 - 7 = \square$	
8.	$11 - 7 = \square$		23.	$16 - 7 = \square$	
9.	$13 - 7 = \square$		24.	$17 - 7 = \square$	
10.	$12 - 9 = \square$		25.	$16 - \square = 9$	
11.	$13 - 9 = \square$		26.	$16 - \square = 8$	
12.	$14 - 9 = \square$		27.	$17 - \square = 8$	
13.	$12 - 8 = \square$		28.	$17 - \square = 9$	
14.	$13 - 8 = \square$		29.	$17 - \square = 16 - 8$	
15.	$14 - 8 = \square$		30.	$\square - 7 = 17 - 8$	

B

Respuestas correctas:

Nombre ______________________ Fecha ______________

*Escribe el número faltante.

1.	10 - 9 = □		16.	11 - 7 = □	
2.	11 - 9 = □		17.	12 - 7 = □	
3.	12 - 9 = □		18.	15 - 7 = □	
4.	10 - 8 = □		19.	15 - 9 = □	
5.	11 - 8 = □		20.	15 - 8 = □	
6.	12 - 8 = □		21.	15 - 7 = □	
7.	10 - 7 = □		22.	15 - 8 = □	
8.	11 - 7 = □		23.	16 - 8 = □	
9.	12 - 7 = □		24.	16 - 7 = □	
10.	11 - 9 = □		25.	16 - □ = 9	
11.	12 - 9 = □		26.	16 - □ = 8	
12.	15 - 9 = □		27.	16 - □ = 7	
13.	11 - 8 = □		28.	16 - □ = 9	
14.	12 - 8 = □		29.	16 - □ = 15 - 8	
15.	15 - 8 = □		30.	□ - 8 = 15 - 7	

A

Respuestas correctas:

Nombre ______________________ Fecha ______________

*Escribe el número faltante.

1.	2 + □ = 3		16.	2 + □ = 8	
2.	1 + □ = 3		17.	4 + □ = 8	
3.	□ + 1 = 3		18.	8 = □ + 6	
4.	□ + 2 = 4		19.	8 = 3 + □	
5.	3 + □ = 4		20.	□ + 3 = 9	
6.	1 + □ = 4		21.	2 + □ = 9	
7.	1 + □ = 5		22.	9 = □ + 1	
8.	4 + □ = 5		23.	9 = 4 + □	
9.	3 + □ = 5		24.	2 + 2 + □ = 9	
10.	3 + □ = 6		25.	2 + 2 + □ = 8	
11.	□ + 2 = 6		26.	3 + □ + 3 = 9	
12.	0 + □ = 6		27.	3 + □ + 2 = 9	
13.	1 + □ = 7		28.	5 + 3 = □ + 4	
14.	□ + 5 = 7		29.	□ + 4 = 1 + 5	
15.	□ + 4 = 7		30.	3 + □ = 2 + 6	

B

Respuestas correctas:

Nombre ______________________ Fecha ____________

*Escribe el número faltante.

1.	$1 + \square = 3$		16.	$3 + \square = 8$	
2.	$0 + \square = 3$		17.	$2 + \square = 8$	
3.	$\square + 3 = 3$		18.	$8 = \square + 1$	
4.	$\square + 2 = 4$		19.	$8 = 4 + \square$	
5.	$3 + \square = 4$		20	$\square + 2 = 9$	
6.	$4 + \square = 4$		21.	$4 + \square = 9$	
7.	$4 + \square = 5$		22.	$9 = \square + 5$	
8.	$1 + \square = 5$		23.	$9 = 6 + \square$	
9.	$2 + \square = 5$		24.	$1 + 5 + \square = 9$	
10.	$4 + \square = 6$		25.	$3 + 2 + \square = 8$	
11.	$\square + 2 = 6$		26.	$2 + \square + 6 = 9$	
12.	$3 + \square = 6$		27.	$3 + \square + 4 = 9$	
13.	$3 + \square = 7$		28.	$5 + 4 = \square + 6$	
14.	$\square + 4 = 7$		29.	$\square + 3 = 6 + 2$	
15.	$\square + 5 = 7$		30.	$4 + \square = 2 + 7$	

A

Respuestas correctas:

Nombre ______________________ Fecha ____________

*Escribe el número faltante.

1.	2 + □ = 3		16.	2 + □ = 8	
2.	1 + □ = 3		17.	4 + □ = 8	
3.	□ + 1 = 3		18.	8 = □ + 6	
4.	□ + 2 = 4		19.	8 = 3 + □	
5.	3 + □ = 4		20.	□ + 3 = 9	
6.	1 + □ = 4		21.	2 + □ = 9	
7.	1 + □ = 5		22.	9 = □ + 1	
8.	4 + □ = 5		23.	9 = 4 + □	
9.	3 + □ = 5		24.	2 + 2 + □ = 9	
10.	3 + □ = 6		25.	2 + 2 + □ = 8	
11.	□ + 2 = 6		26.	3 + □ + 3 = 9	
12.	0 + □ = 6		27.	3 + □ + 2 = 9	
13.	1 + □ = 7		28.	5 + 3 = □ + 4	
14.	□ + 5 = 7		29.	□ + 4 = 1 + 5	
15.	□ + 4 = 7		30.	3 + □ = 2 + 6	

B

Respuestas correctas:

Nombre ______________________ Fecha ______________

*Escribe el número faltante.

1.	$1 + \square = 3$		16.	$3 + \square = 8$	
2.	$0 + \square = 3$		17.	$2 + \square = 8$	
3.	$\square + 3 = 3$		18.	$8 = \square + 1$	
4.	$\square + 2 = 4$		19.	$8 = 4 + \square$	
5.	$3 + \square = 4$		20.	$\square + 2 = 9$	
6.	$4 + \square = 4$		21.	$4 + \square = 9$	
7.	$4 + \square = 5$		22.	$9 = \square + 5$	
8.	$1 + \square = 5$		23.	$9 = 6 + \square$	
9.	$2 + \square = 5$		24.	$1 + 5 + \square = 9$	
10.	$4 + \square = 6$		25.	$3 + 2 + \square = 8$	
11.	$\square + 2 = 6$		26.	$2 + \square + 6 = 9$	
12.	$3 + \square = 6$		27.	$3 + \square + 4 = 9$	
13.	$3 + \square = 7$		28.	$5 + 4 = \square + 6$	
14.	$\square + 4 = 7$		29.	$\square + 3 = 6 + 2$	
15.	$\square + 5 = 7$		30.	$4 + \square = 2 + 7$	

A

Respuestas correctas:

Nombre ______________________ Fecha ____________

*Escribe el número faltante.

1.	2 - □ = 1		16.	6 - □ = 2	
2.	2 - □ = 2		17.	6 - □ = 3	
3.	2 - □ = 0		18.	6 - □ = 4	
4.	3 - □ = 2		19.	7 - □ = 3	
5.	3 - □ = 1		20.	7 - □ = 2	
6.	3 - □ = 0		21.	7 - □ = 1	
7.	3 - □ = 3		22.	8 - □ = 2	
8.	4 - □ = 4		23.	8 - □ = 3	
9.	4 - □ = 3		24.	4 = 8 - □	
10.	4 - □ = 2		25.	2 = 9 - □	
11.	4 - □ = 1		26.	3 = 9 - □	
12.	5 - □ = 0		27.	4 = 9 - □	
13.	5 - □ = 1		28.	10 - 3 = 9 - □	
14.	5 - □ = 2		29.	9 - □ = 10 - 5	
15.	5 - □ = 3		30.	9 - □ = 10 - 6	

B

Respuestas correctas:

Nombre ______________________ Fecha ____________

*Escribe el número faltante.

1.	2 - □ = 2		16.	6 - □ = 3	
2.	2 - □ = 1		17.	6 - □ = 4	
3.	2 - □ = 0		18.	6 - □ = 5	
4.	3 - □ = 3		19.	7 - □ = 4	
5.	3 - □ = 2		20.	7 - □ = 3	
6.	3 - □ = 1		21.	7 - □ = 2	
7.	3 - □ = 0		22.	8 - □ = 3	
8.	4 - □ = 4		23.	8 - □ = 4	
9.	4 - □ = 3		24.	5 = 8 - □	
10.	4 - □ = 2		25.	3 = 9 - □	
11.	4 - □ = 1		26.	4 = 9 - □	
12.	5 - □ = 5		27.	5 = 9 - □	
13.	5 - □ = 4		28.	10 - 4 = 9 - □	
14.	5 - □ = 3		29.	9 - □ = 10 - 6	
15.	5 - □ = 2		30.	9 - □ = 10 - 5	

A

Respuestas correctas:

Nombre ______________________ Fecha ______________

*Escribe el número faltante.

1.	□ = 4 + 1		16.	7 + 3 = 4 + □	
2.	□ = 4 + 2		17.	6 + 4 = 5 + □	
3.	□ = 4 + 3		18.	5 + 5 = 6 + □	
4.	□ = 5 + 1		19.	5 + 3 = □ + 1	
5.	□ = 5 + 2		20.	5 + 4 = □ + 5	
6.	□ = 5 + 3		21.	4 + 5 = □ + 5	
7.	□ = 6 + 1		22.	2 + □ = 6 + 2	
8.	8 = 7 + □		23.	4 + □ = 5 + 3	
9.	9 = 8 + □		24.	□ + 4 = 5 + 2	
10.	9 = □ + 1		25.	□ + 6 = 4 + 3	
11.	9 = □ + 9		26.	4 + 2 = 1 + □	
12.	8 = □ + 1		27.	3 + 4 = □ + 2	
13.	□ = 7 + 1		28.	4 + 4 = 2 + □	
14.	10 = 8 + □		29.	3 + □ = 2 + 7	
15.	10 = □ + 8		30.	□ + 2 = 2 + 6	

B

Respuestas correctas:

Nombre ______________________ Fecha ____________

*Escribe el número faltante.

1.	□ = 3 + 1		16.	5 + 5 = 4 + □	
2.	□ = 3 + 2		17.	6 + 4 = 7 + □	
3.	□ = 3 + 3		18.	3 + 7 = 8 + □	
4.	□ = 4 + 1		19.	5 + 2 = □ + 1	
5.	□ = 4 + 2		20.	5 + 3 = □ + 5	
6.	□ = 4 + 3		21.	4 + 4 = □ + 4	
7.	□ = 5 + 1		22.	3 + □ = 6 + 3	
8.	8 = 1 + □		23.	4 + □ = 5 + 4	
9.	9 = 1 + □		24.	□ + 4 = 2 + 5	
10.	8 = □ + 7		25.	□ + 6 = 3 + 4	
11.	8 = □ + 8		26.	4 + 3 = 1 + □	
12.	7 = □ + 1		27.	4 + 4 = □ + 2	
13.	□ = 6 + 1		28.	4 + 5 = 2 + □	
14.	10 = 9 + □		29.	3 + □ = 2 + 6	
15.	10 = □ + 9		30.	□ + 2 = 2 + 7	

A

Respuestas correctas:

Nombre ______________________ Fecha ____________

*Escribe el número faltante.

1.	10 + 3 = □		16.	10 + □ = 11	
2.	10 + 2 = □		17.	10 + □ = 12	
3.	10 + 1 = □		18.	5 + □ = 15	
4.	1 + 10 = □		19.	4 + □ = 14	
5.	4 + 10 = □		20.	□ + 10 = 17	
6.	6 + 10 = □		21.	17 − □ = 7	
7.	10 + 7 = □		22.	16 − □ = 6	
8.	8 + 10 = □		23.	18 − □ = 8	
9.	12 − 10 = □		24.	□ − 10 = 8	
10.	11 − 10 = □		25.	□ − 10 = 9	
11.	10 − 10 = □		26.	1 + 1 + 10 = □	
12.	13 − 10 = □		27.	2 + 2 + 10 = □	
13.	14 − 10 = □		28.	2 + 3 + 10 = □	
14.	15 − 10 = □		29.	4 + □ + 3 = 17	
15.	18 − 10 = □		30.	□ + 5 + 10 = 18	

B

Respuestas correctas:

Nombre ______________________ Fecha ______________

*Escribe el número faltante.

1.	10 + 1 = □		16.	10 + □ = 10	
2.	10 + 2 = □		17.	10 + □ = 11	
3.	10 + 3 = □		18.	2 + □ = 12	
4.	4 + 10 = □		19.	3 + □ = 13	
5.	5 + 10 = □		20.	□ + 10 = 13	
6.	6 + 10 = □		21.	13 – □ = 3	
7.	10 + 8 = □		22.	14 – □ = 4	
8.	8 + 10 = □		23.	16 – □ = 6	
9.	10 – 10 = □		24.	□ – 10 = 6	
10.	11 – 10 = □		25.	□ – 10 = 8	
11.	12 – 10 = □		26.	2 + 1 + 10 = □	
12.	13 – 10 = □		27.	3 + 2 + 10 = □	
13.	15 – 10 = □		28.	2 + 3 + 10 = □	
14.	17 – 10 = □		29.	4 + □ + 4 = 18	
15.	19 – 10 = □		30.	□ + 6 + 10 = 19	

A

Respuestas correctas:

Nombre ______________________ Fecha ______________

*Escribe el número faltante.

1.	10 + 2 = □		16.	12 + 3 = □	
2.	2 + 1 = □		17.	13 + 3 = □	
3.	10 + 3 = □		18.	14 + 3 = □	
4.	4 + 10 = □		19.	13 + 5 = □	
5.	4 + 2 = □		20.	14 + 5 = □	
6.	6 + 10 = □		21.	15 + 5 = □	
7.	10 + 3 = □		22.	4 + 14 = □	
8.	3 + 3 = □		23.	4 + 15 = □	
9.	10 + 6 = □		24.	12 + □ = 14	
10.	2 + 1 = □		25.	12 + □ = 15	
11.	12 + 1 = □		26.	12 + □ = 16	
12.	2 + 2 = □		27.	□ + 4 = 16	
13.	12 + 2 = □		28.	5 + □ = 16	
14.	3 + 3 = □		29.	5 + □ = 26	
15.	13 + 3 = □		30.	4 + □ = 36	

B

Respuestas correctas:

Nombre ______________________ Fecha ____________

*Escribe el número faltante.

1.	10 + 1 = □		16.	12 + 2 = □	
2.	1 + 1 = □		17.	13 + 2 = □	
3.	10 + 2 = □		18.	14 + 2 = □	
4.	3 + 10 = □		19.	13 + 4 = □	
5.	3 + 2 = □		20.	14 + 4 = □	
6.	5 + 10 = □		21.	15 + 4 = □	
7.	10 + 2 = □		22.	5 + 14 = □	
8.	2 + 2 = □		23.	5 + 15 = □	
9.	10 + 4 = □		24.	11 + □ = 12	
10.	2 + 1 = □		25.	11 + □ = 13	
11.	12 + 1 = □		26.	11 + □ = 14	
12.	1 + 1 = □		27.	□ + 3 = 14	
13.	11 + 1 = □		28.	6 + □ = 19	
14.	3 + 2 = □		29.	6 + □ = 29	
15.	13 + 2 = □		30.	5 + □ = 39	

1.er grado
Módulo 3

A

Respuestas correctas:

Nombre ______________________ Fecha ______________

*Escribe el número faltante.

1.	$3 - 3 = \square$		16.	$13 - 1 = \square$	
2.	$13 - 3 = \square$		17.	$13 - 2 = \square$	
3.	$3 - 2 = \square$		18.	$14 - 3 = \square$	
4.	$13 - 2 = \square$		19.	$14 - 4 = \square$	
5.	$4 - 2 = \square$		20.	$14 - 10 = \square$	
6.	$14 - 2 = \square$		21.	$17 - 5 = \square$	
7.	$4 - 3 = \square$		22.	$17 - 6 = \square$	
8.	$14 - 3 = \square$		23.	$17 - 10 = \square$	
9.	$14 - 10 = \square$		24.	$8 - \square = 5$	
10.	$7 - 6 = \square$		25.	$18 - \square = 15$	
11.	$17 - 6 = \square$		26.	$18 - \square = 13$	
12.	$17 - 10 = \square$		27.	$19 - \square = 12$	
13.	$6 - 3 = \square$		28.	$\square - 2 = 17$	
14.	$16 - 3 = \square$		29.	$17 - 3 = 16 - \square$	
15.	$16 - 10 = \square$		30.	$19 - 6 = \square - 5$	

B

Respuestas correctas:

Nombre ______________________ Fecha ______________

*Escribe el número faltante.

1.	$2 - 2 = \square$		16.	$14 - 1 = \square$	
2.	$12 - 2 = \square$		17.	$14 - 2 = \square$	
3.	$2 - 1 = \square$		18.	$15 - 3 = \square$	
4.	$12 - 1 = \square$		19.	$15 - 4 = \square$	
5.	$3 - 3 = \square$		20.	$15 - 10 = \square$	
6.	$13 - 3 = \square$		21.	$18 - 5 = \square$	
7.	$3 - 2 = \square$		22.	$18 - 6 = \square$	
8.	$13 - 2 = \square$		23.	$18 - 10 = \square$	
9.	$13 - 10 = \square$		24.	$7 - \square = 5$	
10.	$6 - 5 = \square$		25.	$17 - \square = 15$	
11.	$16 - 5 = \square$		26.	$17 - \square = 13$	
12.	$16 - 10 = \square$		27.	$19 - \square = 13$	
13.	$4 - 2 = \square$		28.	$\square - 3 = 16$	
14.	$14 - 2 = \square$		29.	$17 - 4 = 16 - \square$	
15.	$14 - 10 = \square$		30.	$19 - 7 = \square - 6$	

A

Respuestas correctas:

Nombre ______________________ Fecha ______________

*Escribe el número faltante. Presta atención a los signos de + y –.

1.	5 + 2 = □		16.	13 + 6 = □	
2.	15 + 2 = □		17.	3 + 16 = □	
3.	2 + 5 = □		18.	19 – 2 = □	
4.	12 + 5 = □		19.	19 – 7 = □	
5.	7 – 2 = □		20.	4 + 15 = □	
6.	17 – 2 = □		21.	14 + 5 = □	
7.	7 – 5 = □		22.	18 – 6 = □	
8.	17 – 5 = □		23.	18 – 2 = □	
9.	4 + 3 = □		24.	13 + □ = 19	
10.	14 + 3 = □		25.	□ – 6 = 13	
11.	3 + 4 = □		26.	14 + □ = 19	
12.	13 + 4 = □		27.	□ – 4 = 15	
13.	7 – 4 = □		28.	□ – 5 = 14	
14.	17 – 4 = □		29.	13 + 4 = 19 – □	
15.	17 – 3 = □		30.	18 – 6 = □ + 3	

B

Respuestas correctas:

Nombre ______________________ Fecha ____________

*Escribe el número faltante. Presta atención a los signos de + y –.

1.	5 + 1 = □		16.	12 + 7 = □	
2.	15 + 1 = □		17.	2 + 17 = □	
3.	1 + 5 = □		18.	18 – 2 = □	
4.	11 + 5 = □		19.	18 – 6 = □	
5.	6 – 1 = □		20.	3 + 16 = □	
6.	16 – 1 = □		21.	13 + 6 = □	
7.	6 – 5 = □		22.	17 – 4 = □	
8.	16 – 5 = □		23.	17 – 3 = □	
9.	4 + 5 = □		24.	12 + □ = 18	
10.	14 + 5 = □		25.	□ – 6 = 12	
11.	5 + 4 = □		26.	13 + □ = 19	
12.	15 + 4 = □		27.	□ – 3 = 16	
13.	9 – 4 = □		28.	□ – 3 = 17	
14.	19 – 4 = □		29.	11 + 6 = 19 – □	
15.	19 – 5 = □		30.	19 – 5 = □ + 3	

A

Respuestas correctas:

Nombre ______________________ Fecha ____________

*Escribe el número faltante.

1.	17 - 1 = □		16.	19 - 9 = □	
2.	15 - 1 = □		17.	18 - 9 = □	
3.	19 - 1 = □		18.	11 - 9 = □	
4.	15 - 2 = □		19.	16 - 5 = □	
5.	17 - 2 = □		20.	15 - 5 = □	
6.	18 - 2 = □		21.	14 - 5 = □	
7.	18 - 3 = □		22.	12 - 5 = □	
8.	18 - 5 = □		23.	12 - 6 = □	
9.	17 - 5 = □		24.	14 - □ = 11	
10.	19 - 5 = □		25.	14 - □ = 10	
11.	17 - 7 = □		26.	14 - □ = 9	
12.	18 - 7 = □		27.	15 - □ = 9	
13.	19 - 7 = □		28.	□ - 7 = 9	
14.	19 - 2 = □		29.	19 - 5 = 16 - □	
15.	19 - 7 = □		30.	15 - 8 = □ - 9	

B

Respuestas correctas:

Nombre ____________________ Fecha ____________

*Escribe el número faltante.

1.	16 - 1 = □		16.	19 - 9 = □	
2.	14 - 1 = □		17.	18 - 9 = □	
3.	18 - 1 = □		18.	12 - 9 = □	
4.	19 - 2 = □		19.	19 - 8 = □	
5.	17 - 2 = □		20.	18 - 8 = □	
6.	15 - 2 = □		21.	17 - 8 = □	
7.	15 - 3 = □		22.	14 - 5 = □	
8.	17 - 5 = □		23.	13 - 5 = □	
9.	19 - 5 = □		24.	12 - □ = 7	
10.	16 - 5 = □		25.	16 - □ = 10	
11.	16 - 6 = □		26.	16 - □ = 9	
12.	19 - 6 = □		27.	17 - □ = 9	
13.	17 - 6 = □		28.	□ - 7 = 9	
14.	17 - 1 = □		29.	19 - 4 = 17 - □	
15.	17 - 6 = □		30.	16 - 8 = □ - 9	

A

Respuestas correctas:

Nombre ______________________ Fecha ______________

*Escribe el número faltante.

1.	17 + 1 = □		16.	11 + 9 = □	
2.	15 + 1 = □		17.	10 + 9 = □	
3.	18 + 1 = □		18.	9 + 9 = □	
4.	15 + 2 = □		19.	7 + 9 = □	
5.	17 + 2 = □		20.	8 + 8 = □	
6.	18 + 2 = □		21.	7 + 8 = □	
7.	15 + 3 = □		22.	8 + 5 = □	
8.	5 + 13 = □		23.	11 + 8 = □	
9.	15 + 2 = □		24.	12 + □ = 17	
10.	5 + 12 = □		25.	14 + □ = 17	
11.	12 + 4 = □		26.	8 + □ = 17	
12.	13 + 4 = □		27.	□ + 7 = 16	
13.	3 + 14 = □		28.	□ + 7 = 15	
14.	17 + 2 = □		29.	9 + 5 = 10 + □	
15.	12 + 7 = □		30.	7 + 8 = □ + 9	

B

Respuestas correctas:

Nombre ______________________ Fecha ______________

*Escribe el número faltante.

1.	14 + 1 = □		16.	11 + 9 = □	
2.	16 + 1 = □		17.	10 + 9 = □	
3.	17 + 1 = □		18.	8 + 9 = □	
4.	11 + 2 = □		19.	9 + 9 = □	
5.	15 + 2 = □		20.	9 + 8 = □	
6.	17 + 2 = □		21.	8 + 8 = □	
7.	15 + 4 = □		22.	8 + 5 = □	
8.	4 + 15 = □		23.	11 + 7 = □	
9.	15 + 3 = □		24.	12 + □ = 18	
10.	5 + 13 = □		25.	14 + □ = 18	
11.	13 + 4 = □		26.	8 + □ = 18	
12.	14 + 4 = □		27.	□ + 5 = 14	
13.	4 + 14 = □		28.	□ + 6 = 15	
14.	16 + 3 = □		29.	9 + 6 = 10 + □	
15.	13 + 6 = □		30.	6 + 7 = □ + 9	

A

Respuestas correctas:

Nombre ______________________ Fecha ____________

*Escribe el número faltante.

1.	17 + 1 = □		16.	11 + 9 = □	
2.	15 + 1 = □		17.	10 + 9 = □	
3.	18 + 1 = □		18.	9 + 9 = □	
4.	15 + 2 = □		19.	7 + 9 = □	
5.	17 + 2 = □		20.	8 + 8 = □	
6.	18 + 2 = □		21.	7 + 8 = □	
7.	15 + 3 = □		22.	8 + 5 = □	
8.	5 + 13 = □		23.	11 + 8 = □	
9.	15 + 2 = □		24.	12 + □ = 17	
10.	5 + 12 = □		25.	14 + □ = 17	
11.	12 + 4 = □		26.	8 + □ = 17	
12.	13 + 4 = □		27.	□ + 7 = 16	
13.	3 + 14 = □		28.	□ + 7 = 15	
14.	17 + 2 = □		29.	9 + 5 = 10 + □	
15.	12 + 7 = □		30.	7 + 8 = □ + 9	

B

Respuestas correctas:

Nombre ____________________ Fecha ____________

*Escribe el número faltante.

1.	14 + 1 = □		16.	11 + 9 = □	
2.	16 + 1 = □		17.	10 + 9 = □	
3.	17 + 1 = □		18.	8 + 9 = □	
4.	11 + 2 = □		19.	9 + 9 = □	
5.	15 + 2 = □		20.	9 + 8 = □	
6.	17 + 2 = □		21.	8 + 8 = □	
7.	15 + 4 = □		22.	8 + 5 = □	
8.	4 + 15 = □		23.	11 + 7 = □	
9.	15 + 3 = □		24.	12 + □ = 18	
10.	5 + 13 = □		25.	14 + □ = 18	
11.	13 + 4 = □		26.	8 + □ = 18	
12.	14 + 4 = □		27.	□ + 5 = 14	
13.	4 + 14 = □		28.	□ + 6 = 15	
14.	16 + 3 = □		29.	9 + 6 = 10 + □	
15.	13 + 6 = □		30.	6 + 7 = □ + 9	

A

Respuestas correctas:

Nombre ______________________ Fecha ____________

*Escribe el número faltante.

1.	17 - 1 = □		16.	19 - 9 = □	
2.	15 - 1 = □		17.	18 - 9 = □	
3.	19 - 1 = □		18.	11 - 9 = □	
4.	15 - 2 = □		19.	16 - 5 = □	
5.	17 - 2 = □		20.	15 - 5 = □	
6.	18 - 2 = □		21.	14 - 5 = □	
7.	18 - 3 = □		22.	12 - 5 = □	
8.	18 - 5 = □		23.	12 - 6 = □	
9.	17 - 5 = □		24.	14 - □ = 11	
10.	19 - 5 = □		25.	14 - □ = 10	
11.	17 - 7 = □		26.	14 - □ = 9	
12.	18 - 7 = □		27.	15 - □ = 9	
13.	19 - 7 = □		28.	□ - 7 = 9	
14	19 - 2 = □		29.	19 - 5 = 16 - □	
15.	19 - 7 = □		30.	15 - 8 = □ - 9	

B

Respuestas correctas:

Nombre ______________________ Fecha ____________

*Escribe el número faltante.

1.	$16 - 1 = \square$		16.	$19 - 9 = \square$	
2.	$14 - 1 = \square$		17.	$18 - 9 = \square$	
3.	$18 - 1 = \square$		18.	$12 - 9 = \square$	
4.	$19 - 2 = \square$		19.	$19 - 8 = \square$	
5.	$17 - 2 = \square$		20.	$18 - 8 = \square$	
6.	$15 - 2 = \square$		21.	$17 - 8 = \square$	
7.	$15 - 3 = \square$		22.	$14 - 5 = \square$	
8.	$17 - 5 = \square$		23.	$13 - 5 = \square$	
9.	$19 - 5 = \square$		24.	$12 - \square = 7$	
10.	$16 - 5 = \square$		25.	$16 - \square = 10$	
11.	$16 - 6 = \square$		26.	$16 - \square = 9$	
12.	$19 - 6 = \square$		27.	$17 - \square = 9$	
13.	$17 - 6 = \square$		28.	$\square - 7 = 9$	
14.	$17 - 1 = \square$		29.	$19 - 4 = 17 - \square$	
15.	$17 - 6 = \square$		30.	$16 - 8 = \square - 9$	

A

Respuestas correctas:

Nombre ______________________ Fecha ____________

*Escribe el número faltante.

1.	9 + 1 + 3 = □		16.	6 + 3 + 8 = □	
2.	9 + 2 + 1 = □		17.	5 + 9 + 4 = □	
3.	5 + 5 + 3 = □		18.	3 + 12 + 4 = □	
4.	5 + 2 + 5 = □		19.	3 + 11 + 5 = □	
5.	4 + 5 + 5 = □		20.	5 + 6 + 7 = □	
6.	8 + 2 + 4 = □		21.	2 + 6 + 3 = □	
7.	8 + 3 + 2 = □		22.	3 + 2 + 13 = □	
8.	12 + 2 + 2 = □		23.	3 + 13 + 3 = □	
9.	3 + 3 + 12 = □		24.	9 + 1 + □ = 14	
10.	4 + 4 + 5 = □		25.	8 + 4 + □ = 16	
11.	2 + 15 + 2 = □		26.	□ + 8 + 6 = 19	
12.	7 + 3 + 3 = □		27.	2 + □ + 7 = 18	
13.	1 + 17 + 1 = □		28.	2 + 2 + □ = 18	
14.	14 + 2 + 2 = □		29.	19 = 6 + □ + 9	
15.	4 + 12 + 4 = □		30.	18 = 7 + □ + 6	

B

Respuestas correctas:

Nombre ______________________________ Fecha ________________

*Escribe el número faltante.

1.	9 + 1 + 2 = □		16.	6 + 3 + 9 = □	
2.	9 + 4 + 1 = □		17.	4 + 9 + 2 = □	
3.	5 + 5 + 1 = □		18.	2 + 12 + 4 = □	
4.	5 + 3 + 5 = □		19.	2 + 11 + 5 = □	
5.	4 + 5 + 5 = □		20.	6 + 6 + 7 = □	
6.	8 + 2 + 2 = □		21.	2 + 6 + 5 = □	
7.	8 + 3 + 2 = □		22.	3 + 3 + 13 = □	
8.	11 + 1 + 1 = □		23.	3 + 14 + 3 = □	
9.	2 + 2 + 14 = □		24.	9 + 1 + □ = 13	
10.	4 + 4 + 4 = □		25.	8 + 4 + □ = 15	
11.	2 + 13 + 2 = □		26.	□ + 8 + 6 = 18	
12.	6 + 3 + 3 = □		27.	2 + □ + 6 = 18	
13.	1 + 15 + 1 = □		28.	2 + 5 + □ = 18	
14.	15 + 2 + 2 = □		29.	19 = 5 + □ + 9	
15.	3 + 14 + 3 = □		30.	19 = 7 + □ + 6	

Créditos

Great Minds® ha hecho todos los esfuerzos para obtener permisos para la reimpresión de todo el material protegido por derechos de autor. Si algún propietario de material sujeto a derechos de autor no ha sido mencionado, favor ponerse en contacto con Great Minds para su debida mención en todas las ediciones y reimpresiones futuras.